Frontiers in Mathematics

Jorge Bustamante

Algebraic Approximation:

A Guide to Past and Current Solutions

 Birkhäuser

Jorge Bustamante
Facultad de Ciencias Físico-Matemáticas
Benemérita Universidad Autónoma de Puebla
Apartado Postal 1152
Puebla, Pue. C.P. 72000
Mexico
jbusta@fcfm.buap.mx

ISSN 1660-8046   e-ISSN 1660-8054
ISBN 978-3-0348-0193-5   e-ISBN 978-3-0348-0194-2
DOI 10.1007/978-3-0348-0194-2
Springer Basel Dordrecht Heidelberg London New York

Library of Congress Control Number: 2011941776

Mathematics Subject Classification (2010): 41-02, 41-03, 41A25, 41A27, 41A36, 41A50, 41A52

Printed on acid-free paper

Springer Basel AG is part of Springer Science+Business Media
(www.birkhauser-science.com)

Contents

Preface   vii

1   Some Notes on Trigonometric Approximation
   1.1   Early years   1
   1.2   Direct and converse results: a motivation   4
   1.3   Some asymptotic results   9
   1.4   An abstract approach   9

2   The End Points Effect
   2.1   Two different problems   11
   2.2   Nikolskii's discovery   12
   2.3   Problems connected with Nikolskii's result   13
   2.4   Timan-type estimates   16
   2.5   Estimates with higher-order moduli   19
   2.6   Gopengauz-Teliakovskii-type estimates   23
   2.7   Characterization of some classes of functions   27
   2.8   Simultaneous approximation   34
   2.9   Zamansky-type estimates   44
   2.10 Fuksman-Potapov solution to the second problem   46
   2.11 Integral metrics   56
   2.12 L_p, $0 < p < 1$ 62
   2.13 The Whitney theorem   64
   2.14 Other classes of functions   66

3   Looking for New Moduli
   3.1   The works of Potapov   67
   3.2   Butzer and the method of Fourier transforms   73
   3.3   The τ modulus of Ivanov   76
   3.4   Ditzian-Totik moduli   81
       3.4.1   Direct and converse results   82
       3.4.2   Approximation in weighted spaces   85

3.4.3 Marchaud inequalities 86
3.4.4 Simultaneous approximation 87
3.4.5 A Banach space approach 92
3.5 Felten modulus 99

4 Exact Estimates and Asymptotics
4.1 Asymptotics for $\text{Lip}_1(M, [-1, 1]$ 101
4.2 Estimates for W^r 103
4.3 Asymptotics for $C^{r,w}[-1, 1]$ 104
4.4 Estimates for integrable functions 105
4.5 Pointwise asymptotics 107

5 Construction of Special Operators
5.1 Estimates in norm 115
5.2 Timan-type estimates 119
5.3 Gopengauz estimates 124
5.4 Bernstein interpolation process 131
5.4.1 Bernstein first interpolation operators 134
5.4.2 Chebyshev polynomials of second type 134
5.4.3 General Bernstein operators 136
5.4.4 Other modifications 138
5.5 Integral operators 139
5.6 Simultaneous approximation 150
5.7 Estimation with constants 159
5.8 The boolean sums approach 162
5.9 Discrete operators 169

Bibliography 179

Index 205

Preface

This book contains an exposition of several results related with direct and converse theorems in the theory of approximation by algebraic polynomials in a finite interval. In addition, we include some facts concerning trigonometric approximation that are necessary for motivation and comparisons. The selection of papers that we reference and discuss document some trends in polynomial approximation from the 1950s to the present day.

The book does not pretend to be a text for graduate students. We only want to ease the task of understanding the evolution of ideas and to help people in finding the correct references for a specific result. An important feature of the book is to put together some different known solutions to problems in algebraic approximation that are not collected in text books. This explains the large number of references.

Almost all of the material appears in historical order, but the concepts are separated into groups in order to present a fuller picture of the state of the art in a specific problem.

Several topics related with algebraic approximation are not included here. For instance, we do not discuss approximation with constraints, because that would double the length of the book. On the other hand, we do present a few facts concerning approximation by positive linear operators.

I hope that this survey will be helpful to students and researchers interested in approximation by algebraic polynomials. Any suggestions that would help to improve these notes would be welcomed by the author.

Jorge Bustamante González
Benemérita Universidad Autónoma de Puebla
Mexico

Chapter 1

Some Notes on Trigonometric Approximation

1.1   Early years

Let us denote by $(C[a, b]\ C[0, 2\pi])$ the space of all real continuous (2π-periodic) functions f provided with the uniform norm

$$\|f\| = \sup_{x \in [a,b]} |f(x)| \quad \left(= \sup_{x \in [0,2\pi]} |f(x)| \right).$$

By $\mathbb{P}_n$ ($\mathbb{T}_n$) we denote the family of all algebraic (trigonometric) polynomials of degree not greater than n. In 1885 Weierstrass published his famous theorem asserting that, every continuous (periodic) function on a compact interval is the limit in the uniform norm of a sequence of algebraic (trigonometric) polynomials. We shall mention that, almost at the same time, Runge showed that an arbitrary continuous function can be approximated by means of a rational function and that rational functions can be approximated by means of polynomials [318] and [319]. But he did not formulate the result explicitly. In a modern notation, Weierstrass's theorem can be written as follows: for any $f \in C[a, b]$,

$$\lim_{n \to \infty} E_n(f, [a, b]) = 0, \tag{1.1}$$

where

$$E_n(f, [a, b]) = \inf\{ \|f - P\| : P \in \mathbb{P}_n\} \tag{1.2}$$

is called the best approximation of f (by algebraic polynomials) of order n. For trigonometric approximation the best approximation is defined analogously. That is, if $f \in C[0, 2\pi]$, then

$$E_n^*(f) = \inf_{T \in \mathbb{T}_n} \|f - T\|.$$

After Weierstrass, several different proofs of the same result appeared. Among others, there are some due to Lebesgue (1898, who used approximations of a function by broken lines [228]), Lerch (1892 and 1903, who approximated by a polygonal line and then by a Fourier series, [232] and [233]), Volterra (1987, who used ideas very similar to Lerch's, [401]), Borel (1905, [36]), Landau (1908, who used a singular integral [224]), Simon (1918, who modified Landau's ideas in order to approximate by finite sums [341]), de la Vallée-Poussin (1908, who provided an elegant proof in the trigonometrical case using a special integral, see [82] and [85]) and Kryloff (1908, who used a discrete version of de la Vallée-Poussin's integral, [222]). For interesting comments related with these results see the paper of Pinkus [281].

As Jackson remarked [177], *a time came when there was no longer any distinction in inventing a proof of Weierstrass's theorem, unless the new method could be shown to possess some specific excellence.* At that time it was known that there exist some connections between the smoothness of a function and its approximation by partial sums of Fourier series. These ideas can be found in Picard's book [280].

It was Lebesgue in 1908 who formally stated the problem of studying the relation between smoothness and best approximation [229]. He considered the problem for Lipschitz functions. We say that $L \in \text{Lip}_\alpha[a, b]$ ($0 < \alpha \leq 1$), if there exists a constant $K = K(f)$ such that

$$| f(x) - f(y) | \leq K | x - y |^\alpha . \qquad (1.3)$$

We also set

$$\text{Lip}_\alpha(M, [a, b]) = \{f : [a, b] \to \mathbb{R} :| f(x) - f(y) | \leq M | x - y |^\alpha\}. \qquad (1.4)$$

Lebesgue proved that if $f \in \text{Lip}_1[a, b]$, then

$$E_n(f) \leq C\sqrt{(log n)/n}.$$

In [82] de la Vallée-Poussin improved the estimate by showing that

$$E_n(f) \leq C/\sqrt{n}.$$

The study of the special function $g(x) =| x |$, $x \in [-1, 1]$, played an important role. It belongs to $g \in \text{Lip}_1[-1, 1]$. In 1908 de la Vallée-Poussin [81] constructed a polynomial P_n such that

$$| P_n(x)- | x | |< \frac{C}{n}.$$

Two years later he proved that we can not find polynomials P_n satisfying

$$| P_n(x)- | x | |< \frac{C}{n \log^3 n}$$

for all $x \in [-1, 1]$. Bernstein improved this result.

Theorem 1.1.1 (Bernstein, [25]). *Let C and ε be positive numbers and $r \in \mathbb{N}$. If $f \in C[a,b]$ and for each $n \in \mathbb{N}$ there exists $P_n \in \mathbb{P}_n$ such that*

$$| P_n(x) - f(x) | < \frac{C}{n^r (\log n)^{1+\varepsilon}};$$

then $f \in C^r[a,b]$.

Thus for the case of $g(x) = | x |$ ($x \in [-1,1]$) and $r = 1$ such an inequality is not possible. The correct estimate for this function was found by Bernstein in his prize essay for the Belgian Academy [26]:

$$E_n(g) \geq C/\sqrt{n}.$$

One important point in the Bernstein paper quoted above is that, for the first time, there appears what now is called Bernstein's inequality: if $T_n \in \mathbb{T}_n$, then

$$\|T_n'\| \leq n \|T_n\|. \tag{1.5}$$

In fact, in the original paper Bernstein proved that $\|T_n'\| \leq 2n \|T_n\|$. The inequality in the form (1.5) was presented by de la Vallée-Poussin in [84]. For an algebraic polynomial the inequality can be stated as: if $P_n \in \mathbb{P}_n$, then

$$| \sqrt{1 - x^2} P_n'(x) | \leq n \|P_n\|_\infty, \qquad x \in [-1,1]. \tag{1.6}$$

Bernstein considered first the algebraic case, but Jackson [178] noticed that it is simpler to study first the trigonometrical case.

The relevance of Bernstein's inequality comes from its applications to converse results. As an example we recall here one of the assertions obtained in this way. If $f \in C[0, 2\pi]$ and $E_n^*(f) \leq C/n^{k+\alpha}$ ($0 < \alpha < 1$), then f has a continuous kth derivative and $f^{(k)} \in \mathrm{Lip}_\alpha[0, 2\pi]$.

Concerning the direct result, Jackson in his dissertation and in [176] proved that, if a function f of period 2π satisfies condition (1.3) (with $\alpha = 1$), then

$$E_n^*(f) \leq \frac{CK}{n},$$

where C is an absolute constant, $\pi/2 \leq C \leq 3$. If $f : [a,b] \to \mathbb{R}$ satisfies condition (1.3), then

$$E_n(f) \leq \frac{CK(b-a)}{n},$$

where C is an absolute constant, $1/2 \leq C \leq 3/2$.

1.2   Direct and converse results: a motivation

Since this work is devoted to approximation by algebraic polynomials, we omit many details related with the history of trigonometric approximation.

In Theorem 1.2.1 we present a summary of the best results concerning direct and converse results and then we give some historical remarks.

For each $r \in \mathbb{N}$ and $f \in L_p[0, 2\pi]$ $(1 \le p \le \infty)$ the usual modulus of continuity of order r is defined by

$$\omega_r(f, t)_p = \sup_{h \in (0, t]} \|\Delta_h^r f\|_p,$$

where $\Delta_h^r f(x) = \sum_{k=0}^r (-1)^k \binom{r}{k} f(x + kh)$ is the central difference of order r with step h. For the case of continuous functions we omit the index p in the above notation and when $r = 1$ we also omit the index r.

We say that $\omega : [0, a] \to \mathbb{R}$ is a modulus of continuity, if ω is an increasing continuous function $\omega(0) = 0$, $\omega(t) > 0$ for $t > 0$, and

$$\omega(t + s) \le \omega(s) + \omega(t). \tag{1.7}$$

Theorem 1.2.1. *Fix $f \in C[0, 2\pi]$, $r, s \in \mathbb{N}_0$ and σ such that $r < \sigma < s$ and let $\{T_n\}$ be the sequence of polynomials of the best approximation for f. The following assertions are equivalent:*

(i) $\qquad\qquad E_n^*(f) = \|f - T_n\| = \mathcal{O}(n^{-\sigma}), \qquad\qquad (n \to \infty),$

(ii) $\qquad\qquad\qquad \omega_s(f, t) = \mathcal{O}(t^\sigma), \qquad\qquad\qquad (t \to 0),$

(iii) $\qquad\qquad\qquad \|T_n^{(s)}\| = \mathcal{O}(n^{-(\sigma - s)}), \qquad\qquad (n \to \infty),$

(iv) $\quad f \in C^r[0, 2\pi]$ and $\|f^{(r)} - T_n^{(r)}\| = \mathcal{O}(n^{-(\sigma - r)}), \qquad (n \to \infty),$

(v) $\quad f \in C^r[0, 2\pi]$ and $\omega_1(f^{(r)}, t) = \mathcal{O}(t^{\sigma - r}), \qquad (0 < \sigma - r < 1),$

(vi) $\quad f \in C^r[0, 2\pi]$ and $\omega_2(f^{(r)}, t) = \mathcal{O}(t^{\sigma - r}), \qquad (0 < \sigma - r < 2).$

The fact that the assertions in Theorem 1.2.1 are equivalent not only in $C[0, 2\pi]$ but in the setting of normed spaces was proved by Butzer and Scherer ([51] and [52]). They showed that essentially what we need is to have on hand appropriate Jackson and Bernstein-type inequalities. The abstract approach will be presented in the last section of this chapter.

The assertion (v) $\Rightarrow$ (i) was proved by Jackson [175] and [176] (the statement presented here is not the original). Different versions were later developed by Favard [115], Akhieser and Krein [1] and Korneichuk [207].

Theorem 1.2.2 (Direct result). *For each $r \in \mathbb{N}$, $f \in C^r[0, 2\pi]$ and $n \in \mathbb{N}_0$,*

$$E_n^*(f) \le \frac{C(r)}{n^r} \omega\left(f^{(r)}, \frac{1}{n}\right) \qquad and \qquad E_n^*(f) \le \frac{K_r}{n^{r+1}} \|f^{(r)}\|,$$

where K_r is Favard's constant defined by

$$K_r = \frac{4}{\pi} \sum_{k=0}^{\infty} \frac{(-1)^{k(r+1)}}{(2k+1)^{r+1}}. \tag{1.8}$$

It is known that $K_0 = 1$, $K_1 = \pi/2$, $K_2 = \pi^2/8$, $K_3 = \pi^3/24$ and

$$\frac{\pi^2}{8} = K_2 < K_4 < \cdots < \frac{4}{\pi} < \cdots < K_3 < K_1 = \frac{\pi}{2}.$$

In 1912 Bernstein studied the converse result. The proof of Bernstein is a model for almost all converse theorems and was based on Bernstein's inequality

$$\|T_n^{(r)}\| \leq n^r \, \|T_n\| \qquad (T_n \in \mathbb{T}_n),$$

and $n \in \mathbb{N}$) which follows from (1.5) by induction.

Let $\mathcal{W}[0, 2\pi]$ be the class of all functions $f \in C[0, 2\pi]$ for which there exists a constant $C = C(f)$ such that

$$\omega(f, t) \leq C\, t(1 + |\ln t|).$$

Theorem 1.2.3 (Bernstein). *Fix $\alpha \in (0, 1]$ and $f \in C[0, 2\pi]$ and suppose there exists a constant C such that $E_n^*(f) \leq cn^{-\alpha}$. Then, if $\alpha < 1$, $f \in \mathrm{Lip}_\alpha[0, 2\pi]$ and if $\alpha = 1$, $f \in \mathcal{W}[0, 2\pi]$.*

In 1919 de la Vallée-Poussin [85] (following Bernstein's ideas) proved the following theorem.

Theorem 1.2.4. *Let $\Omega : [a, \infty)$ $(a > 1)$ be a decreasing function such that*

$$\lim_{t \to \infty} \Omega(t) = 0 \quad and \quad \int_a^\infty \frac{\Omega(u)}{u} du < \infty.$$

If $p \in \mathbb{N}$, $f \in C[0, 2\pi]$ and $E_n^*(f) \leq \Omega(n)n^{-p}$, then $f^{(p)}$ exists and

$$\omega(f^{(p)}, t) \leq C \left(t \int_a^{a/t} \Omega(u) du + \int_{1/t}^\infty \frac{\Omega(u)}{u} du \right). \tag{1.9}$$

Following Freud, here $\mathcal{O}$ may not be substituted by o. Thus if, for some $r \geq 0$ and $0 < \alpha < 1$, there is a polynomial $T_n \in \mathbb{T}_n$ such that for $n \in \mathbb{N}$ we have

$$\|f - T_n\| \leq \frac{C}{n^{r+\alpha}},$$

then $f \in C^r[0, 2\pi]$ and $f^{(r)} \in \mathrm{Lip}_\alpha[0, 2\pi]$. That is (i) $\Rightarrow$ (v) (for $\alpha \neq 1$). Thus the problem of the characterization of the class of functions with an rth derivative in $\mathrm{Lip}_\alpha[0, 2\pi]$ was completely solved for the case $0 < \alpha < 1$. The case $\alpha = 1$ is not included in the last results. Bernstein proved that condition $E_n^*(f) = O(n^{-1})$ does not imply $f \in \mathrm{Lip}_1[0, 2\pi]$. In particular the function $f(x) = \sum_{k=1}^\infty k^{-2} \sin(kx)$ provides a counterexample.

Zygmund showed that we must consider a wider class:

$$Z[0, 2\pi] = \{f : C[0, 2\pi] \to \mathbb{R} \; : \; | \Delta_h^2 f(x) | \leq Ah, \; h \in (0, 1] \}.$$

Theorem 1.2.5 (Zygmund, [420]). *If $f \in C[0, 2\pi]$, $r \in \mathbb{N}$ and $0 < \alpha < 2$, then $E_n^*(f) = \mathcal{O}(n^{-r-\alpha})$ if and only if $| \Delta_h^2 f^{(r)}(x) | \leq Ch^\alpha$.*

Thus the assertion (i) $\Leftrightarrow$ (vi) for $\sigma - r = 1$ is proved. In [420] Zygmund also included some results related with non-periodic functions. Montel [253] studied several facts concerning the class Z.

For $1 < p < \infty$ Timan [382] proved a sharper version of the Jackson-type inequality for the best trigonometric approximation:

$$n^{-r} \left(\sum_{k=1}^{r} k^{sr-1} E_k^*(f)_p^s \right)^{1/s} \leq C(r, p) \omega_r(f, 1/n)_p, \tag{1.10}$$

where $s = \max\{p, 2\}$.

The converse inequality is given in the following form:

Theorem 1.2.6 (Timan, [381]). *For $1 < p < \infty$, $q = \min\{p, 2\}$ and $f \in L_p[0, 2\pi]$, one has*

$$\omega_r(f, 1/n)_p \leq C(r, p) n^{-r} \left(\sum_{k=1}^{r} k^{rq-1} E_k^*(f)_p^q \right)^{1/q}.$$

There is also an equivalent relation.

Theorem 1.2.7 (Zygmund, [421]). *For $1 < p < \infty$, $q = \min\{p, 2\}$ and $f \in L_p[0, 2\pi]$, one has*

$$\omega_r(f, t)_p \leq C(r, p) \, t^r \left(\int_t^{1/2} \frac{\omega_{r+1}(f, u)}{u^{qr+1}} du \right)^{1/q}.$$

This last relation is sometimes called a sharp Marchaud inequality. Some extensions were given by Ditzian [96]

In 1949 Zamansky showed that (i) $\Rightarrow$ (iii).

Theorem 1.2.8 (Zamansky, [415]). *Let φ be a positive strictly increasing or decreasing continuous function and fix $f \in C[0, 2\pi]$ and $m \in \mathbb{N}$. Suppose that for each $n \in \mathbb{N}$ there is a polynomial $T_n \in \mathbb{T}_n$ such that*

$$\|f - T_n\| \leq n^{1-m} \varphi(n),$$

then there are constants C_1, C_2 and C_3 such that

$$\|T_n^{(m)}\| \leq C_1 + C_2 n \varphi(n) + C_3 \int_1^n \varphi(u) du.$$

In particular if $\|f - T_n\| = \mathcal{O}(n^{-\beta})$ for $\beta > 0$, then $\|T_n^{(m)}\| = \mathcal{O}(n^{m-\beta})$ with $\beta < m$.

The assertion (i) $\Rightarrow$ (iv) is due to Stechkin.

Theorem 1.2.9 (Stechkin, [347]). *Fix $k \in \mathbb{N}$ and let $\{F_n\}$ be a non-increasing sequence of non-negative numbers such that $\sum_{n=1}^{\infty} n^{k-1} F_n < \infty$. Let $f \in C[0, 2\pi]$ and $\{T_n\}$ $(T_n \in \mathbb{T})$ a sequence such that*

$$\|f - T_n\| \le F_{n+1}, \qquad , n \in \mathbb{N}.$$

Then $f \in C^k[0, 2\pi]$ and there exists a constant C such that

$$\|f^{(k)} - T_n^{(k)}\| \le C \left(n^k F_{n+1} + \sum_{j=n+1}^{\infty} j^{k-1} F_j \right),$$

for each $n \in \mathbb{N}$.

In 1967–1968 Butzer-Pawelke [46] and Sunouchi [360] proved the assertion (iii) $\Rightarrow$ (i). In fact Butzer and Pawelke considered the problem for $L_2[0, 2\pi]$ and Sunouchi for all $L_p[0, 2\pi]$ spaces and $C[0, 2\pi]$.

Theorem 1.2.10. *Fix $m \in \mathbb{N}$, $\beta \in (0, m)$ and $f \in C[0, 2\pi]$ and let $\{T_n\}$ be the sequence of polynomials of the best approximation for f. If $\|T_n^{(m)}\| = \mathcal{O}(n^{m-\beta})$, then $\|f - T_n\| = \mathcal{O}(n^{-\beta})$.*

The results recalled above also hold in $L_p[0, 2\pi]$ spaces $(1 \le p < \infty)$. For instance, for $0 < \alpha < r$,

$$E_n(f)_p = \mathcal{O}(n^{-\alpha}) \iff \omega_k(f, t)_p = \mathcal{O}(t^\alpha). \tag{1.11}$$

We can interpret the equivalence (1.11) in two different forms.

a) The classical Lipschitz spaces are characterized in terms of the best trigonometric approximation.

b) A class of functions with a given rate for the best trigonometric approximation is characterized in terms of Lipschitz classes.

For trigonometric approximation both assertions are the same. In algebraic approximation the situation is different.

Some results in algebraic approximation were obtained by reduction to the trigonometric case. If $f \in C[-1, 1]$, with the change of variable $x \mapsto \cos \theta$ we obtain an even 2π-periodic function $g(\theta) = f(\cos \theta)$. If $P_n \in \mathbb{P}_n$ is the polynomial of best approximation for f, then $T_n(\theta) = P_n(\cos \theta)$ is the trigonometric polynomial of best approximation for g. Therefore, if $f \in C^1[-1, 1]$, then

$$E_n(f) = E_n(g) \le \frac{\pi}{2(n+1)} \|g'\| \le \frac{\pi}{2(n+1)} \|f'\|,$$

where we have used the relations $g'(\cos \theta) = f'(\cos \theta) \sin \theta$ and $| \sin \theta | \le 1$. Thus the precision has been changed. In this way, some theorems related the approximation of non-periodic functions by algebraic polynomials were obtained. For instance, for every $f \in C[a,b]$ and $n \in \mathbb{N}$ there exists a polynomial $P_n \in \mathbb{P}_n$ such that, for all $x \in [a,b]$,

$$| f(x) - P_n(x) | \leq C\,\omega(f, (b-a)/n) \tag{1.12}$$

([179], p. 15) and, if $f \in C^r[a,b]$, then for each $n \in \mathbb{N}$ $(n > r)$ there exists a polynomial $P_n \in \mathbb{P}_n$ such that, for all $x \in [a,b]$,

$$| f(x) - P_n(x) | \leq \frac{C(b-a)^p}{n^p}\,\omega\left(f^{(p)}, \frac{b-a}{n-r}\right),$$

([179], p. 18). In both cases the constant C does not depend upon f or n.

The theory of best approximation of functions by algebraic polynomials was not as complete as in the trigonometric case. A characterization of the class $\mathrm{Lip}_\alpha[-1,1]$ $(0 < \alpha \leq 1)$ in terms of the best approximation was not known. We notice that (1.12) can be written in the more precise form

$$E_n(f) \leq 12\omega\left(f, \frac{b-a}{2n}\right).$$

For the converse results Bernstein can not obtain an analogous form of Theorem 1.2.3. He only found properties on proper subintervals.

Theorem 1.2.11 (Bernstein). *Fix $\alpha \in (0,1]$ and $f \in C[a,b]$ and suppose there exists a constant C such that $E_n(x) \leq cn^{-\alpha}$. Then for each couple of numbers c and d satisfying $a < c < d < b$ one has $f \in \mathrm{Lip}_\alpha[c,d]$ (if $\alpha < 1$) and $f \in \mathcal{W}[c,d]$ (if $\alpha = 1$).*

The restriction to proper subintervals of $[a,b]$ is essential. It was known that for the function $f(x) = \sqrt{1-x^2}$ one has $E_n(f, [-1,1]) < 2/(\pi n)$, while $f \notin \mathrm{Lip}_\alpha[-1,1]$ for any $\alpha > 1/2$. Thus a result like Theorem 1.2.4 holds only on proper subintervals of $[a,b]$.

Some years later, in 1956, Csibi proved that we can obtain a conclusion for the full interval if the rate of approximation is faster.

Theorem 1.2.12 (Csibi, [77]). *Let Ω be as in Theorem 1.2.4. If $p \in \mathbb{N}$, $f \in C[a,b]$ and $E_n(f, [a,b]) \leq \Omega(n^2)n^{-2p}$, then $f^{(p)}$ exists and $\omega(f^{(p)}, t)$ satisfies (1.9).*

Bernstein [30] also considered the case of a higher rate of convergence of the best approximation. In particular, he proved that, if for each $r \in \mathbb{N}$,

$$\lim_{n \to \infty} n^r E_n(f, [a,b]) = 0,$$

then f has derivatives of all order in (a,b).

1.3 Some asymptotic results

There are two typical problems: to estimate the error for a fixed class of function (see (1.15)) and to consider the asymptotic of error in the class (see (1.16)). For the first problem, fix a bounded set $M \subset C[0, 2\pi]$, $n \in \mathbb{N}$ and define

$$E_n^*(M) = \sup_{f \in M} E_n^*(f). \tag{1.13}$$

Of course, we can obtain the exact value of $E_n^*(M)$ only for some special sets M. Even more, it is not easy to know if the sup is attained in some element of M. The analysis of this problem was motivated by some results of Favard and Akhieser-Krein. Set

$$W^{r+1}(M, [0, 2\pi]) = \{f : f^{(r)} \in A.C.[0, 2\pi], \|f^{(r+1)}\|_\infty \leq M\}. \tag{1.14}$$

Favard [115] and Akhieser-Krein [1] proved that

$$E_n^*(W^r(1, [0, 2\pi])) = \frac{K_r}{n^r}, \tag{1.15}$$

where K_r is Favard's constant (1.8), and Nikolskii noticed in [270] that there exist functions $f \in W^r(1, [0, 2\pi])$ for which

$$\lim_{n \to \infty} \sup \, n^r \, E_n^*(f) = K_r. \tag{1.16}$$

1.4 An abstract approach

Let us present an abstract approach introduced by Butzer and Scherer in [51], [52] and [53] (see also [185]).

Let X be a normed space with norm $\|\cdot\|_X$ and $\{M_n\}$ be a sequence of linear subspaces of X such that $M_n \subset M_{n+1}$ and $\cup_{n=1}^{\infty} M_n$ is dense in X. For $f \in X$, the best approximation of f by elements of M_n is defined as

$$E_n(f) = \inf_{g \in M_n} \|f - g\|_X.$$

We assume that, for each $f \in X$ and $n \in \mathbb{N}$, there exists $g_n = g_n(f) \in M_n$ such that $E_n(f) = \|f - g_n\|_X$.

Let Y be a linear subspace of X with a seminorm $|\cdot|_Y$ such that, for each $n \in \mathbb{N}$, $M_n \subset Y$. Set

$$K(f, t; X, Y) = \inf_{g \in Y} \{\|f - g\|_X + t \mid g \mid_Y\}, \qquad (t > 0, f \in X).$$

Theorem 1.4.1. *Let X be a normed linear space and $\{M_n\}$ be a sequence of linear subspaces of X such that $M_n \subset M_{n+1}$ and $\cup_{n=1}^{\infty} M_n$ is dense in X. Fix two linear* *subspaces of X, Y and Z, with a seminorms $|\cdot|_Y$ and $|\cdot|_Z$ respectively such that, for each $n \in \mathbb{N}$, $M_n \subset Y$ and $M_n \subset Z$.*

Fix two real numbers ρ and σ ($0 \le \sigma < \rho$) and assume there exist constants A, B, C and D such that

$$E_n(g) \le \frac{A}{n^\rho} \mid g \mid_Y, \quad \mid g_n \mid_Y \le B\, n^\rho \, \|g_n\|, \qquad (g \in Y, g_n \in M_n, n \in \mathbb{N}), \quad (1.17)$$

$$E_n(h) \le \frac{C}{n^\sigma} \mid h \mid_Z, \quad \mid g_n \mid_Z \le D\, n^\sigma \, \|g_n\| \qquad (h \in Z, g_n \in M_n, n \in \mathbb{N}). \quad (1.18)$$

If Z is a Banach space with the norm $\|\cdot\|_Z = \|\cdot\|_X + |\cdot|_Z$, then the following assertions are equivalent for $f \in X$ and $\sigma < s < \rho$

  (i) $E_n(f) = \mathcal{O}(n^{-s})$, *as $n \to \infty$.*

  (ii) *If $E_n(f) = \|f - g_n\|_X$, then $\mid g_n \mid_Y = \mathcal{O}(n^{\rho-s})$, as $n \to \infty$.*

  (iii) *If $f \in Z$ and $E_n(f) = \|f - g_n\|_X$, then $\mid f - g_n \mid_Z = \mathcal{O}(n^{\sigma-s})$, as $n \to \infty$.*

  (iv) $K(f, t^\rho, X, Y) = \mathcal{O}(t^s)$, *as $t \to 0$.*

Chapter 2

The End Points Effect

2.1 Two different problems

As we have noticed, the theorems related with direct and converse results for trigonometric approximation can not be translated word by word to the case of algebraic approximation. Thus we have two different questions:

1. Given a modulus of smoothness, how can the associated (generalized) Lipschitz classes be characterized with the help of approximation by means of algebraic polynomials?

2. How can the class of functions with a given rate of algebraic polynomial approximation (say $E_n(f) \leq M/n^\alpha$) be characterized in terms of properties related with smoothness and/or differentiability?

Some solutions to the first problem were given by Nikolskii, Timan and Dzyadyk. They considered the space of continuous functions and uniform approximation. In this approach the use of the quantity

$$\Delta_n(x) = \frac{\sqrt{1-x^2}}{n} + \frac{1}{n^2} \tag{2.1}$$

was essential. Fuksman presented the first results related with solutions of the second problem. Fuksman obtained a characterization of functions $f \in C[-1,1]$ for which $E_n(f) \leq M/n^\alpha$ $(0 < \alpha < 1)$ with the help of a local modulus of continuity.

After the works of Timan some different problems were considered:

(i) When can $\Delta_n(x)$ be changed by

$$\delta_n(x) = \frac{\sqrt{1-x^2}}{n}? \tag{2.2}$$

(ii) When can the polynomials used in approximating functions also approximate the derivatives?

(iii) When can the estimates given in terms of the first modulus of continuity be improved by using higher-order moduli?

(iv) Is it possible to obtain estimates which combine (i) and (ii) (or other relations)?

2.2   Nikolskii's discovery

Since the known estimates for trigonometric approximation did not always lead to optimal results in algebraic approximation, some new methods were needed.

In 1946 Nikolskii made an important advance by considering point-wise estimation by means of a sequence of linear operators. Let $\{T_n\}$ be the sequence of Chebyshev polynomials

$$T_n(x) = \cos(n \arccos x).$$

It is known that these polynomials are orthogonal with respect to the measure $2dt/\pi\sqrt{1-t^2}$ on the interval $[-1,1]$.

As usual, for $f \in C[-1,1]$ the Fourier-Chebyshev coefficients are defined by

$$a_n(f) = \frac{2}{\pi} \int_{-1}^{1} \frac{f(t)T_n(t)dt}{\sqrt{1-t^2}}.$$

For $f \in C[-1,1]$ and $x \in [-1,1]$ define

$$U_n(f,x) = \frac{a_0(f)}{2} + \sum_{k=1}^{n} k\lambda_{n,k}\, a_k(f)\, T_k(x)$$

where

$$\lambda_{n,k} = \frac{\pi}{2n}\, \cot \frac{k\pi}{2n}.$$

Theorem 2.2.1 (Nikolskii, [271]). *For each $n \in \mathbb{N}$, one has $U_n : C[-1,1] \to \mathbb{P}_n$. Moreover, if $f \in \mathrm{Lip}_1(M,[-1,1])$ (see (1.4)) and $n \in \mathbb{N}$, then*

$$| f(x) - U_n(f,x) | \leq \frac{M\pi}{2}\, \frac{\sqrt{1-x^2}}{n+1} + | x |\, \mathcal{O}\left(\frac{\log n}{n^2}\right) \tag{2.3}$$

and $\mathcal{O}$ can not be replaced by o.

Proof. It is sufficient to consider the case $M = 1$. Notice that every function in $\mathrm{Lip}_1[-1,1]$ is absolutely continuous. Let us write

$$K(t) = \sum_{k=1}^{\infty} \frac{\sin kt}{t}, \qquad I_n = \frac{2}{\pi} \int_{0}^{\pi} \left| K(t) - \sum_{k=1}^{n-1} \lambda_{n,k} \sin(kt) \right| dt$$

and

$$J_n = \frac{2}{\pi} \int_0^\pi \left| K(t) - \sum_{k=1}^{n-1} \lambda_{n,k} \sin(kt) \right| \sin t \, dt.$$

By setting $x = \cos\theta$ and integrating by parts, one has

$$f(\cos\theta) = \frac{a_0}{2} + \frac{1}{\pi} \int_0^{2\pi} K(t) \sin(t+\theta) f'(\cos(t+\theta)) dt$$

and

$$U_n(f,\cos\theta) = \frac{a_0}{2} + \frac{1}{\pi} \int_0^{2\pi} \left(\sum_{k=1}^{n-1} \lambda_{n,k} \sin(k+t) \right) \sin(t+\theta) f'(\cos(t+\theta)) dt.$$

Since $\mid f'(\cos(t+\theta)) \mid \le 1$,

$$\mid f(\cos\theta) - U_n(f,\cos\theta) \mid \le \frac{1}{\pi} \int_0^{2\pi} \left| K(t) - \sum_{k=1}^{n-1} \sin(k+t) \right| \mid \sin(t+\theta) \mid dt.$$

$$\le I_n \mid \sin\theta \mid + J_n \mid \cos\theta \mid \le I_n \sqrt{1-x^2} + J_n \mid x \mid.$$

Then Nikolskii proved that $I_n = \pi/(2n)$ and $J_n = \mathcal{O}(\ln n/n^2)$.   □

There is a great difference with Jackson's theorem: the position of x on the interval $[-1,1]$ is taken into account in the factor $\sqrt{1-x^2}$.

Timan and Dzyadik [380] proved that, if $f \in C^r[a,b]$ and $f^{(r)}$ is quasi-smooth (Zygmund), then $E_n(f) = \mathcal{O}(n^{-(r+1)})$. The sentence improves a result of Montel for $E_n(f)$ which gave an estimate only inside of the interval.

2.3   Problems connected with Nikolskii's result

Nikolskii's result motivated several investigations on the possibility of approximation (including the asymptotically best approximation) of functions of various classes by algebraic polynomials and many results concerning the improvement of approximation at the endpoints of the segment $[-1,1]$.

In 1958 Lebed [226] gave an extension of Nikolskii's theorem by considering functions in the Zygmund class:

$$Z[-1,1] = \{ f : C[-1,1] \to \mathbb{R} \; : \mid \Delta_h^2 f(x) \mid \le Mh, \; h \in (0,1] \}.$$

Theorem 2.3.1 (Lebed, [226]). *If $f \in C^m[-1,1]$ and $f^{(m)} \in Z[-1,1]$ (with constant M), then there exists a sequence $\{P_n\}$ ($P_n \in \mathbb{P}_n$) such that*

$$\mid f(x) - P_n(x) \mid \le C(m) M (\Delta_n(x))^m \left(\Delta_n(x) + \frac{\log n}{n^2} \right),$$

where

$$\Delta_n(x) = \left(\sqrt{1-x^2} + \mid x \mid /n \right). \tag{2.4}$$

The factor $\pi/2$ in (2.3) cannot be replaced by a smaller one (see (4.2)). Temlyakov proved the existence of a sequence $\{P_n\}$ for which an estimate with specific constants in both terms holds. His construction was not obtained by means of a sequence of linear operators, but he strengthened inequality (2.3) by omitting $\log n$ in the remainder.

Theorem 2.3.2 (Temlyakov, [372]). *Assume $f \in \mathrm{Lip}_1(1, [-1, 1])$. For any natural number n there exists an algebraic polynomial P_n of degree n such that*

$$| f(x) - P_n(x) | \le \frac{\pi \sqrt{1 - x^2}}{2n} + \frac{\pi^2 \, | x \, |}{8n^2}. \tag{2.5}$$

Proof. The proof is based on an inequality for the best trigonometric approximation of a differentiable function:

$$E_n(h) \le \frac{K_r}{(n+1)^r} E_n(h^{(r)}), \qquad r \in \mathbb{N}, \tag{2.6}$$

where K_r is the Favard constant. Since $K_1 = \pi/2$, what we need is a good representation of the function $g(t) = f(\cos t)$. If

$$-f'(\cos t) = \frac{a_0}{2} + \sum_{k=1}^{\infty} a_k \cos(kt) \qquad \text{and} \qquad \varphi(t) = \sum_{k=1}^{\infty} \frac{a_k}{k} \sin(kt),$$

then g can be written as

$$g(t) = -\frac{a_0}{2} \cos t + \varphi(t) \sin t + \sigma(t) \cos t + G(t),$$

where

$$\sigma(t) = \int_0^t \varphi(s) ds \qquad \text{and} \qquad G'(t) = -\sigma(t) \sin t.$$

Now, let u_{n-1} and v_{n-1} be the trigonometric polynomial of best approximation of order $n-1$ of the functions φ and σ respectively and define $P_n(\cos t) = u_{n-1}(t) \sin t$ and $Q_n(\cos t) = -v_{n-1}(t) \cos t$. Since $E_0(\varphi') \le 1$, it follows from (2.6) that

$$| \varphi(t) - P_n(\cos t) | \le | \varphi(t) - u_{n-1}(t) | | \sin t | \le E_{n-1}(\varphi) | \sin t |$$

$$\le \frac{\pi}{2n} E_{n-1}(\varphi') | \sin t | \le \frac{\pi}{2n} | \sin t |$$

and

$$| \sigma(t) \cos t - Q_n(\cos t) | \le | \sigma(t) - v_{n-1}(t) | | \cos t | \le E_{n-1}(\sigma) | \cos t |$$

$$\le \frac{K_2}{n^2} E_{n-1}(\varphi') | \cos t | \le \frac{K_2}{n^2} | \cos t | \, .$$

Finally, since

$$\frac{d}{dx} G(\arccos x) = \frac{-G'(\arccos x)}{\sqrt{1 - x^2}} = \frac{-G'(t)}{\sin t} = \sigma(t) = \sigma(\arccos x),$$

there exists $S_n \in \mathbb{P}_n$ such that

$$\mid (G(\arccos x))' - S_n(x) \mid \leq E_n(\sigma) \leq \frac{K_2}{(n+1)^2} E_n(\varphi') \leq \frac{K_2}{(n+1)^2}.$$

Thus, if we define

$$R_n(x) = G(\pi/2) + \int_0^x S_n(y)dy$$

then

$$\mid G(\arccos x) - R_n(x) \mid = \left| \int_0^x ((G(\arccos y))' - S_n(y))dy \right| \leq \frac{K_2 \mid x \mid}{(n+1)^2}.$$

The proof finishes by considering the polynomial

$$P(x) = -a_0 x/2 + P_n(x) + Q_n(x) + R_n(x). \qquad \square$$

The second term in (2.5) was written as $\mathcal{O}(\mid x \mid /n^2)$ in the original statement, but we prefer to present here what was really proved. If we want to compare (2.5) with (2.3), for $f \in \mathrm{Lip}_1(M, [-1,1])$ the last term in (2.5) should be multiplied by M. It could be avoided, if such a term can be replaced by zero. But Temlyakov did not know whether such a term can be removed. However, in the same paper he proved the following assertion. For each natural number n one can find a function $f_n \in \mathrm{Lip}_1(1, [-1,1]$ for which there exists no polynomial $P_n \in \mathbb{P}_n$ such that

$$\mid f_n(x) - P_n(x) \mid \leq \frac{\pi\sqrt{1-x^2}}{2(n+1)}.$$

From Theorem 2.3.2, making use of some arguments of Teliakovskii [371], Temliakov obtained the following theorem (the constant in $\mathcal{O}(1/n)$ was not given).

Theorem 2.3.3 (Temlyakov, [372]). *Let $f \in \mathrm{Lip}_1(1, [-1,1])$. For any natural number n there exists an algebraic polynomial P_n of degree n such that*

$$\mid f(x) - P_n(x) \mid \leq \frac{\pi\sqrt{1-x^2}}{2(n+1)} (1 + \mathcal{O}(1/n)).$$

In the chapter devoted to asymptotics we will include some other results. For differentiable functions Ligun presented in 1980 a version which provides some information concerning the constants.

Theorem 2.3.4 (Ligun, [236]). *Let r be an odd number. Then for any function $f \in C^r[-1,1]$ there exists a sequence of algebraic polynomials $\{Q_{n,r}(x)\}$ of degree not greater than $n \geq r$ such that, uniformly with respect to $x \in [-1,1]$,*

$$\mid f(x) - Q_{n-1,r}(x) \mid \leq \frac{K_r(\delta_n(x))^r}{2} \omega\left(f^{(r)}, \pi\delta_n(x)\right) + \mathcal{O}\left(\frac{1}{n^r}\omega\left(f^{(r)}, \frac{1}{n}\right)\right).$$

The proof of the last theorem is very long and technical, so it will not be included here. But we notice that the construction was obtained by means of linear operators.

For a given modulus of continuity w and $r \in \mathbb{N}$, the associated Lipschitz class is defined by

$$H_\omega^k = \{f \in C[a,b] : \omega_k(f,t) \le C(f)\omega(t)\}.$$

Theorem 2.3.5 (Polovina, [286]). *Let w be a modulus of continuity. For each function $f \in C^1[-1,1]$ such that $f' \in H_w$, there exists a sequence of polynomials $\{P_n(f,x)\}$ $(P_n \in \mathbb{P}_n)$ such that*

$$\mid f(x) - P_{n-1}(x) \mid \le \frac{1}{2} \int_0^{\pi\sqrt{1-x^2}/n} w(t)dt + o\left(\frac{1}{n}w\left(\frac{1}{n}\right)\right).$$

Moreover, if w is a concave modulus of continuity, $1/2$ can be changed to $1/4$. The constant $1/4$ cannot be made smaller.

In [18] Bashmakova presented a similar result for functions f such that $f' \in H_w$, for a continuous and concave modulus of continuity.

2.4   Timan-type estimates

In [373] Timan proved that, if $f \in \mathrm{Lip}_\alpha(M, [-1,1])$ and $S_n(f,x)$ is the nth partial sum of the Fourier-Chebyshev series of f, then

$$\mid f(x) - S_n(f,x) \mid \le \frac{2^{\alpha+1}M(1-x^2)^{\alpha/2}\log n}{\pi^2} \frac{1}{n^2} \int_0^{\pi/2} t^\alpha \sin t\, dt + \mathcal{O}\left(\frac{1}{n^\alpha}\right).$$

Later, in 1951, he [374] improved the Nikolskii estimate as follows: for

$$f \in \mathrm{Lip}_\alpha([-1,1]) \quad (0 < \alpha \le 1)$$

one can find a sequence $\{P_n\}$ $(P_n \in \mathbb{P}_n)$ such that

$$\mid f(x) - P_n(x) \mid \le \frac{C}{n^\alpha}\left((\sqrt{1-x^2})^\alpha + \left(\frac{\mid x \mid}{n}\right)^\alpha\right). \tag{2.7}$$

In the same year he generalized this result. For the proof we need the Jackson (also called Jackson-Matsuoka) kernels. [250]

$$K_{n,s}(t) = c_{n,s}\left(\frac{\sin(nt/2)}{\sin(t/2)}\right)^{2s}, \tag{2.8}$$

where $c_{n,s}$ is chosen from the condition $\pi^{-1}\int_{-\pi}^{\pi} K_{n,s}(t)dt = 1$.

Theorem 2.4.1 (Timan, [375]). *For $r \in \mathbb{N}_0$ there exists a constant C_r such that, for each $f \in C^r[-1, 1]$ and $n \in \mathbb{N}$, one can find a polynomial $P_n(f) \in \mathbb{P}_n$ satisfying*

$$| f(x) - P_n(f, x) | \leq C_r \left(\frac{\sqrt{1 - x^2}}{n} + \frac{1}{n^2} \right)^r \omega \left(f^{(r)}, \frac{\sqrt{1 - x^2}}{n} + \frac{1}{n^2} \right). \qquad (2.9)$$

Proof. The proof presented in [379] (p. 262–266) follows by an inductive argument with respect to r. Here we show the case $r = 0$. Define

$$Q_{2n-2}(f, x) = \frac{1}{\pi} \int_0^\pi f(\cos t) \left[K_{n,2}(t + y) + K_{n,2}(t - y) \right] dt,$$

where $x = \cos y$. It can be proved that $Q_{2n-2}(f) \in \mathbb{P}_{2n-2}$. On the other hand,

$$| f(x) - Q_{2n-2}(f, x) | = \left| \frac{1}{\pi} \int_0^\pi [f(\cos y) - f(\cos t)] \left[K_{n,2}(t + y) + K_{n,2}(t - y) \right] dt \right|$$

$$\leq \frac{1}{\pi} \int_0^\pi \omega(f, | \cos y - \cos t |) \left[K_{n,2}(t + y) + K_{n,2}(t - y) \right] dt$$

$$= \frac{1}{2\pi} \int_{-\pi}^\pi \omega \left(f, 2 \left| \sin \frac{t + y}{2} \sin \frac{t - y}{2} \right| \right) \left[K_{n,2}(t + y) + K_{n,2}(t - y) \right] dt.$$

Let us estimate the integral corresponding to $K_{n,2}(t + y)$ (the other one can be estimated with similar arguments).

$$\frac{1}{2\pi} \int_{-\pi}^\pi \omega \left(f, 2 \left| \sin \frac{t + y}{2} \sin \frac{t - y}{2} \right| \right) K_{n,2}(t + y) dt$$

$$= \frac{1}{\pi} \int_{-\pi/2}^{\pi/2} \omega \left(f, 2 \left| \sin(t) \sin(t + y) \right| \right) K_{n,2}(2t) dt$$

$$\leq \frac{1}{\pi} \int_0^{\pi/2} [\omega(f, t^2) + \omega(f, t \mid \sin y \mid)] K_{n,2}(2t) dt.$$

If we consider that

$$\omega(f, t^2) \leq (1 + n^2 t^2) \omega(f, 1/n^2)$$

and

$$\omega(f, t \mid \sin y \mid)) \leq (1 + nt) \omega(f, \sqrt{1 - x^2}/n),$$

it is sufficient to verify that there exists a constant C (independent of n) such that

$$\int_0^{\pi/2} [1 + nt + (nt)^2] K_{n,2}(2t) dt \leq C. \qquad \square$$

In particular, if $\| f^{(r)} \| \leq M$,

$$| f(x) - P_n(x) | \leq \frac{M C_r}{n^r} \left(\sqrt{1 - x^2} + \frac{| x |}{n} \right)^r. \qquad (2.10)$$

Theorem 2.4.2 (Timan [376]). *If w is a modulus of continuity for which*

$$\sum_{n=1}^{\infty} \frac{1}{n} w\left(\frac{1}{n}\right) < \infty$$

and if, for $f \in C[-1,1]$ and algebraic polynomials P_n of degree at most n, $n = 1, 2, 3, \ldots$,

$$\mid f(x) - P_n(x) \mid \leq \Delta_n(x) w(\Delta_n(x)),$$

then $f \in C^1[-1,1]$.

Some years later Hasson showed that this last theorem cannot be improved.

Theorem 2.4.3 (Hasson, [157]). *Let $\{a_n\}$ be an increasing sequence of positive numbers such that $\sum_{n=1}^{\infty} 1/na_n = \infty$, then there exists a function f defined on $[0,1]$ and not continuously differentiable on that interval such that $E_n(f) = \mathcal{O}(1/na_n)$.*

Theorem 2.4.4 ([157]). *Let $f \in C[0,1]$. If $\sum_{n=1}^{\infty} n^{2r-1} E_n(f) < \infty$, then $f \in C^r[0,1]$, $r \in N$.*

Theorem 2.4.5 ([157]). *For every positive integer k and for every $0 < \alpha < 1$, there exists a function $f \in C[0,1]$ such that, for $n \in \mathbb{N}$, $E_n(f) \leq C_1 n^{-2(k+\alpha)}$ and such that $C_2 n^{-\alpha} \leq \omega(f^{(k)}, 1/n) \leq C_2 n^{-\alpha}$.*

Corollary 2.4.6 ([157]). *For every positive integer r and for every $0 < \beta < 1$, there exists a function $f \in C[0,1]$, $f \notin C^r[0,1]$ and $E_n(f) = \mathcal{O}(n^{2r-\beta})$.*

In 1958 Timan noticed that some asymptotics can be improved, if we take into account the position of the point x on the interval $[-1,1]$. From (2.10) we know that, if $f \in W^r(M, [-1,1])$ (see (1.14)) and $x \in [-1,1]$, then

$$\lim_{n\to\infty} \sup n^r \mid f(x) - P_n(f,x) \mid \leq M \, C_r (\sqrt{1-x^2})^r.$$

As Timan proved in [378], instead of C_r we can take Favard's constant

$$\lim_{n\to\infty} \sup n^r \mid f(x) - P_n(f,x) \mid \leq M \, K_r (\sqrt{1-x^2})^r \qquad (2.11)$$

and K_r is the best constant for this kind of inequality. The construction of Timan was connected with the asymptotic best linear method of approximation in the class $W^r(M, [-1,1])$. He suggested that the same idea can be used to construct other asymptotic best linear methods and he considered some method of summation of Fourier series.

Theorem 2.4.1 involves the first modulus of continuity. In the note [377] of 1957, Timan also extended the Nikoslkii estimate (2.3) to the case of functions in the Zygmund class. For $f \in Z[-1,1]$ he constructed a sequence $\{P_n\}$ such that

$$\mid f(x) - P_n(x) \mid \leq C \, \Delta_n(x).$$

The results of Timan motivated investigations in several directions that we will present below. Some authors tried to change the first modulus of continuity in (2.9) by moduli of higher order. Brudnyi showed that such a change is possible. Others looked for results in which the function $\Delta_n(x)$ is replaced by $\delta_n(x)$. This approach began with the works of Teliakovskii and Gopengauz. On the other hand, Trigub noticed that the results of Timan can be generalized to include simultaneous approximation. That is, the same polynomials are used to approximate the functions and several derivatives. Finally, we can consider combinations of these ideas.

2.5   Estimates with higher-order moduli

In 1958 Dzyadyk [109] constructed some kernels which allowed him to give a new (and simpler) proof of Theorem 2.4.1 For a fixed $k \in \mathbb{N}$, $x \in [-\sqrt{2}, \sqrt{2}]$ and $n \in \mathbb{N}$, he defined

$$D_{n,k}(x) = \frac{1}{\gamma_{n,k}} \left(\frac{\sin \frac{1}{2} n \arccos\left(1 - x^2/2\right)}{\sin \frac{1}{2} \arccos\left(1 - x^2/2\right)} \right)^{2k}, \tag{2.12}$$

where $\gamma_{n,k}$ is taken from the condition $\int_{-1}^{1} D_{n,k}(x)dx = 1$. It is an even positive (in $[-\sqrt{2}, \sqrt{2}]$) algebraic polynomial of degree $2k(n-1)$ which can be written in terms of the Chebyshev polynomials T_n in the form

$$(D_{n,k}(x))^{1/k} = \gamma_{n,1} D_{n,1}(x) = 2\frac{1 - T_n(1 - x^2/2)}{x^2}.$$

Using these kernels, one can transform different approximation results related with the Féjer kernels to results concerning approximation by algebraic polynomials. In fact, with the substitution $x = 2\sin(t/2)$ we obtain from an even trigonometric kernel an algebraic kernel with similar properties in the neighborhood of the origin, and vice versa.

Dzyadyk not only presented a new proof of the direct result, he also improved (2.9) by using the second-order modulus instead of the first one. In fact, the kernel constructed in [109] is only used to provided a new proof of Theorem 2.4.1. The assertion relative to the second-order modulus appears (without proof) as a footnote on p. 343.

Theorem 2.5.1 (Dzyakyk, [109]). *For each $r \in \mathbb{N}_0$, there exists a constant C_r such that, for each $f \in C^r[a,b]$ and $n \in \mathbb{N}$ we can construct a polynomial $P_n(f) \in \mathbb{P}_n$ such that*

$$\mid f(x) - P_n(f,x) \mid \leq C_r \left(\rho_n(x)\right)^r \omega_2\left(f^{(r)}, \rho_n(x)\right), \tag{2.13}$$

where

$$\rho_n(x) = \frac{\sqrt{(b-x)(x-a)}}{n} + \frac{1}{n^2}.$$

Proof. The proof is presented for the interval $[0,1]$ and $r = 1$.

1) We can assume that $w_2(f,t) > 0$ for $t > 0$ and $f(0) = f(1)$. It can be proved that there is a constant A such that

$$\frac{1}{n^2} \leq Aw_2\left(f, \frac{1}{n}\right).$$

2) Set $g(x) = f(1-x)$, $x \in [0,1]$. It can be proved that there are extensions F and G of the functions f and g to the interval $[0,4]$ such that

$$w_2(F,t) \leq 25w_2(f,t) \quad \text{and} \quad w_2(G,t) \leq 25w_2(g,t).$$

3) Define

$$\varphi(x) = 2\int_0^{1/3} F(x^2 + 9u^2)D_{n,3}(u)du$$

and

$$\psi(x) = 2\int_0^{1/3} F(x^2 + \frac{9}{2}u^2)D_{n,3}(u)du.$$

Since

$$\left| f(x^2) - 2\psi(x) + \varphi(x) \right|$$

$$= \left| f(x^2)\int_{1/3}^1 D_{n,3}(u)du \right.$$

$$\left. + \int_0^{1/3} \left(F(x^2) - 2F\left(x^2 + \frac{9}{2}u^2\right) + F\left(x^2 + 9u^2\right) \right) D_{n,3}(u)du \right|$$

$$\leq \frac{C_1}{n^5}\|f\| + 2\int_0^{1/3} w_2\left(f, \frac{9}{2}u^2\right) K_{n,3}(u)du$$

$$\leq \frac{C_1}{n^5}\|F\| + 2w_2\left(f, \frac{1}{n^2}\right)\int_0^{1/3}\left(1 + \frac{9}{2}(nu)^2\right)^2 K_{n,3}(u)du \leq C_2 w_2\left(f, \frac{1}{n^2}\right),$$

it is sufficient to approximate φ and ψ with polynomials in x^2.

4) Set

$$P_1(x^2) = \frac{1}{6}\int_{-2}^2 F(u^2)\left[K_{n,3}\left(\frac{u+x}{3}\right) + K_{n,3}\left(\frac{u-x}{3}\right)\right] du$$

$$= \frac{1}{2}\int_{(-2+x)/3}^{(2+x)/3} F((3u-x)^2)K_{n,3}(u)du + \frac{1}{2}\int_{(-2-x)/3}^{(2-x)/3} F((3u+x)^2)K_{n,3}(u)du$$

$$= \frac{1}{2} \int_{-1/3}^{1/3} [F(x^2 - 6xu + 9u^2) + F(x^2 + 6xu + 9u^2)K_{n,3}(u)du$$

$$+ \frac{1}{2} \left(\int_{(-2+x)/3}^{-1/3} + \int_{1/3}^{(2+x)/3} \right) F((3u - x)^2)K_{n,3}(u)du$$

$$+ \frac{1}{2} \left(\int_{(-2-x)/3}^{-1/3} + \int_{1/3}^{(2-x)/3} \right) F((3u - x)^2)K_{n,3}(u)du$$

$$= \frac{1}{2} \int_{-1/3}^{1/3} [F(x^2 - 6xu + 9u^2) + F(x^2 + 6xu + 9u^2)K_{n,3}(u)du + \|F\|\mathcal{O}(n^{-5}).$$

Therefore

$$| \varphi(x) - P_1(x^2) | \leq \int_0^{1/3} \omega_2(F, 6xu)K_{n,3}(u)du + C_3\|F\|\frac{1}{n^4}$$

$$\leq \omega_2 \left(F, \frac{x}{n} \right) \int_0^{1/3} (1 + 6un)^2 K_{n,3}(u)du + C_3\|F\|\frac{1}{n^4}$$

$$\leq C_4 \left(\omega_2 \left(F, \frac{x}{n} \right) + \omega_2 \left(F, \frac{1}{n^2} \right) \right).$$

The analogous construction for ψ is obtained by setting

$$P_2(x^2) = \frac{\sqrt{2}}{6} \int_{-2}^{2} F(u^2) \left[K_{n,3} \left(\sqrt{2}\frac{u + x}{3} \right) + K_{n,3} \left(\sqrt{2}\frac{u - x}{3} \right) \right] du.$$

Thus, if we define

$$P_n(f, x^2) = 2P_2(x^2) - P_1(x^2),$$

then

$$| f(x) - P_n(f, x) | \leq C_5 \left(\omega_2 \left(F, \frac{\sqrt{x}}{n} \right) + \omega_2 \left(F, \frac{1}{n^2} \right) \right)$$

for $x \in [0, 1]$.

If we realize the analogous construction for the function G, then

$$| f(x) - P_n(G, 1 - x) | = | G(1 - x) - P_n(G, 1 - x) |$$

$$\leq C_5 \left(\omega_2 \left(F, \frac{\sqrt{1 - x}}{n} \right) + \omega_2 \left(F, \frac{1}{n^2} \right) \right).$$

For the final construction take $m = [n/3]$ and define

$$P_m(x^2) = (1 - x)P_m(F, x) + xP_m(G, x).$$

We know that $P_m \in \mathbb{P}_n$ and

$$\mid f(x) - P_n(x) \mid \le (1 - x) \mid f(x) - P_n(f, 1 - x) \mid + x \mid f(x) - P_n(G, 1 - x) \mid$$

$$\le C_6 \left((1 - x)\omega_2 \left(F, \frac{\sqrt{x}}{n} \right) + x\omega_2 \left(F, \frac{\sqrt{1 - x}}{n} \right) + \omega_2 \left(F, \frac{1}{n^2} \right) \right)$$

$$\le C_7 \left(\omega_2 \left(F, \frac{\sqrt{x(1 - x)}}{n} \right) + \omega_2 \left(F, \frac{1}{n^2} \right) \right). \qquad \square$$

Dzyadyk also presented the following result.

Theorem 2.5.2 (Dzyakyk, [109]). *Assume that* $f : [-1, 1] \to \mathbb{R}$ *and* $r \in \mathbb{N}_0$. *One has* $f \in C^r[-1, 1]$ *and* $f^{(r)} \in \mathrm{Lip}_1[-1, 1]$ *if and only if there exists a sequence of polynomials* $\{P_n\}$ $(P_n \in \mathbb{P}_n)$, *such that*

$$\mid f(x) - P_n(x) \mid = o\left\{ \left(\frac{\Delta_n(x)}{n} \right)^{r+1} \right\}.$$

In 1959 Freud [123] (independently of Dzyadyk) constructed another sequence of polynomials for which a Timan result holds in terms of the second-order modulus. Freud used the method of intermediate spaces. That is, he first approximates the function f by an adequate piecewise linear function g and then approximates g by polynomials. Freud said that the construction of polynomial kernels (such as the one used by Timan) is not a simple task and he stated the problem of obtaining similar results using differences of higher order and good estimations for the constants. The extension to moduli of smoothness of arbitrary order was given by Brudnyi in 1963.

Theorem 2.5.3 (Brudnyi, [38]). *Given* $r \in \mathbb{N}$, *there exists a constant* C_r *such that, for each* $f \in C[-1, 1]$ *and* $n \in \mathbb{N}$ $(n \ge r - 1)$, *there exists a polynomial* $P_n(f) \in \mathbb{P}_n$ *such that*

$$\mid f(x) - P_n(f, x) \mid \le C_r \omega_r (f, \Delta_n(x)). \qquad (2.14)$$

Let Φ^k denote the class of all non-decreasing continuous functions φ such that, $\varphi(0) = 0$ and $\varphi(t)/t^k$ is non-increasing. Sometimes this last condition is changed by the weaker one: $\varphi(t)/t^k \le C\varphi(s)/s^k$, for $0 < s < t$. Functions of these classes are said to be of the type of the kth order modulus of continuity. It is known that if $\omega_k(f, t) \ne 0$, then $\omega_k(f, t) \in \Phi^k$ (see [247] and [347]).

For $\varphi \in \Phi^k$ and fixed constant M, set

$$H_k^\varphi(M, [-1, 1]) = \{f : [-1, 1] \to \mathbb{R} : \omega_k(f, h) \le M\varphi(h), h \in (0, 1/k]\}$$

and

$$W^r H_k^\varphi(M, [-1, 1]) = \{f; f^{(k)} \in H_k^\varphi(M, [-1, 1])\}$$

$(W^0(M, [-1, 1]) = H_k^\varphi(M, [-1, 1]))$. Moreover set

$$W^r H_k[\varphi] = \bigcup_{M > 0} W^r H_k^\varphi(M, [-1, 1]).$$

Theorem 2.5.4. *Let k and r be natural numbers and assume*

$$\int_0^{b-a} \frac{\omega_{k+r}(f,u)}{u^{r+1}} du < \infty.$$

Then, for $0 < t \leq b - a$,

$$\omega_k(f^{(r)}, t) \leq C(r,k) \int_0^t \frac{\omega_{k+r}(f,u)}{u^{r+1}} du.$$

This theorem was proved by Marchaud in [247] for $k = 1$. For $k \geq 2$ the result was also proved by Brudnyi and Gopengauz in [41].

Guseinov [155] considered the problem of obtaining necessary and sufficient conditions on $\varphi \in \Phi^k$ and $w \in \Phi^{k+r}$ under which the equality $H_w^{k+r} = W^r H_\varphi^k$ holds.

Theorem 2.5.5 (Guseinov, [155]). *Let k, r be natural numbers and $w \in \Phi^r$ and set $\varphi(t) = w(t)/t^r$. Then $H_w^{k+r} = W^r H_\varphi^k$ if and only if*

$$\int_0^t \frac{\omega(u)}{u^{r+1}} du \leq C(r,k) \frac{w(t)}{t^r}.$$

Theorem 2.5.6 (Guseinov, [155]). *Let k, r be natural numbers, $w \in \Phi^{r+k}$ and set $\varphi(t) \in \Phi^k$. Then $H_w^{k+r} = W^r H_\varphi^k$ if and only if $\varphi(t) \leq Cw(t)/t^r$ and*

$$\int_0^t \frac{\omega(u)}{u^{r+1}} du \leq C\varphi(t).$$

For the classes Φ^k the Brudnyi theorem yields

Theorem 2.5.7. *If $\varphi \in \Phi^k$, $r \in \mathbb{N}_0$ and $k \in \mathbb{N}$, there exists a constant $C = (r,k)$ such that, if $f \in W^r H_k^\varphi(1,[-1,1])$, then for any natural number $n \geq r + k - 1$ there is an algebraic polynomial P_n of degree not greater than n, such that*

$$| f(x) - P_n(x) | \leq C(r,k) (\Delta_n(x))^r \varphi(\Delta_n(x)). \tag{2.15}$$

2.6 Gopengauz-Teliakovskii-type estimates

In a Timan-type estimate the term $\Delta_n(x) = \sqrt{1-x^2}/n + 1/n^2$ appears. It is natural to ask whether such a term can be replaced by the simpler one $\delta_n(x) = \sqrt{1-x^2}/n$. A positive answer follows from the works of Teliakovskii (1966) using the first modulus of continuity

Theorem 2.6.1 (Teliakovskii, [371]). *Let r be a non-negative integer. There exists a constant C_r such that, for each $f \in C^r[-1,1]$ and $n > r$ one can find a polynomial $Q_n(f) \in \mathbb{P}_n$ such that*

$$| f(x) - Q_n(f,x) | \leq C_r (\delta_n(x))^r \omega\left(f^{(r)}, \delta_n(x)\right). \tag{2.16}$$

Proof. We will consider only the case $r > 0$. Fix polynomials $P_n(f)$ such that the estimates (2.9) in Timan's Theorem 2.4.1 holds and define

$$Q_n(x) = P_n(f, x) + R_n(f, x)$$

where $R_n(f) \in \mathbb{P}_q$ is the polynomial which interpolates the function $f(x) - P_n(f, x)$ and its derivatives up to the order $q = [r/2]$ at the points ± 1. Teliakovskii proved that Q_n satisfies (2.16). The proof also uses some ideas of simultaneous approximation which will be presented in another section. In particular it follows from Theorem 2.8.10 that

$$\mid f^{(k)}(\pm 1) - P_n(f)^{(k)}(\pm 1) \mid \leq \frac{R}{n^{2(r-k)}} \omega \left(f^{(r)}, \frac{1}{n^2} \right),$$

where the constant R is independent of f and n. Now using the formula of interpolation of Hermite one has

$$\mid R_n(f, x) \mid \leq R \sum_{j=0}^{q} \frac{(1 - x^2)^j}{n^{2(r-j)}} \omega \left(f^{(r)}, \frac{1}{n^2} \right) \tag{2.17}$$

and for $0 \leq k \leq r$ and $x \in [-1, 1]$,

$$\mid R_n^{(k)}(f, x) \mid \leq \frac{C}{n^{2(r-q)}} \omega \left(f^{(r)}, \frac{1}{n^2} \right). \tag{2.18}$$

If $1/n \leq \sqrt{1 - x^2}$, then from (2.9) and (2.17) one has

$$\mid f(x) - Q_n(x) \mid \leq C_r \left(\frac{\sqrt{1 - x^2}}{n} + \frac{1}{n^2} \right)^r \omega \left(f^{(r)}, \frac{\sqrt{1 - x^2}}{n} + \frac{1}{n^2} \right)$$

$$+ R \left(\frac{\sqrt{1 - x^2}}{n} \right)^r \omega \left(f^{(r)}, \frac{1}{n^2} \right) \sum_{j=0}^{q} \frac{1}{(n\sqrt{1 - x^2})^{r-2k}}$$

$$\leq C \left(\delta_n(x) \right)^r \omega \left(f^{(r)}, \delta_n(x) \right).$$

Now assume $1/n \geq \sqrt{1 - x^2}$ and suppose $x > 0$. If $r > 0$ $(q + 1 \leq r)$, then it follows from Theorem 2.8.10 and (2.18) that

$$\mid f(x) - Q_n(x) \mid$$

$$= \left| (-1)^{q+1} \int_x^1 \int_{u_1}^1 \cdots \int_{u_q}^1 [f^{(q+1)}(u) - P_n^{(q+1)}(f, u) - R_n^{(q+1)}(f, u)] du\, du_q \cdots du_1 \right|$$

$$\leq \frac{R}{n^{2(r-q-1)}} \omega \left(f^{(r)}, \frac{1}{n^2} \right) \int_x^1 \int_{u_1}^1 \cdots \int_{u_q}^1 du\, du_q \cdots du_1$$

$$\leq \frac{R}{n^{2(r-q-1)}} \omega \left(f^{(r)}, \frac{1}{n^2} \right) (1 - x^2)^{q+1}.$$

Since $1/n \geq \sqrt{1 - x^2}$,

$$\omega\left(f^{(r)}, \frac{1}{n^2}\right) \leq \frac{2}{n\sqrt{1 - x^2}} \omega\left(f^{(r)}, \frac{\sqrt{1 - x^2}}{n}\right),$$

we obtain

$$|f(x) - Q_n(x)| \leq \frac{R}{n^{2r-2q-1}} (1 - x^2)^{q+1/2} \omega\left(f^{(r)}, \frac{\sqrt{1 - x^2}}{n}\right)$$

$$= R\left(\frac{\sqrt{1 - x^2}}{n}\right)^r \omega\left(f^{(r)}, \frac{\sqrt{1 - x^2}}{n}\right) (n\sqrt{1 - x^2})^{2q+1-r}$$

$$\leq R\left(\frac{\sqrt{1 - x^2}}{n}\right)^r \omega\left(f^{(r)}, \frac{\sqrt{1 - x^2}}{n}\right).$$

For $x < 0$ the result follows analogously.   □

In 1967 Gopengauz obtained a similar result in terms of the modulus of continuity of second order.

Theorem 2.6.2 (Gopengauz, [151]). *For each $f \in C[-1, 1]$ and $n \geq 2$, there exists $P_n \in \mathbb{P}_n$ such that*

$$|f(x) - P_n(x)| \leq C\omega_2(f, \delta_n(x)),$$

where the constant C does not depend on f or n.

Bashmakova and Malozemov gave an estimate including interpolation.

Theorem 2.6.3 (Bashmakova and Malozemov, [17]). *For $f \in C[-1, 1]$ and $-1 = x_0 < x_1 < \cdots < x_n = 1$ $(n > 1)$ there exists $P_n(f) \in \mathbb{P}_n$ such that*

$$|f(x) - P_n(f, x)| \leq A\omega(f, \delta_n(x))$$

and, for $x \in [-1, 1]$ and $k = 1, \ldots, m - 1$,

$$|f(x) - P_n(f, x)| \leq A\omega\left(f, |x - x_k|, \sqrt{|x - x_k|}/n\right).$$

There is also a more complicated version of Theorem 2.6.2 for fractional derivatives.

Theorem 2.6.4 (Shalashova, [336]). *If $f \in C[0, 1]$ has continuous derivatives of fractional order r $(r = r' + \alpha$, with r' integer and $\alpha \in (0, 1))$, there there exists for any $n \geq r - 1$ a polynomial $P_n \in \mathbb{P}_n$ such that*

$$|f(x) - P_n(x) - Ax^r| \leq C_r\left(\frac{\sqrt{x(1 - x)}}{n}\right)^r \omega\left(f^{(r)}, \frac{\sqrt{x(1 - x)}}{n}\right),$$

where C_r does not depend on f or n and A depends on f and r. For fractional r, the term Ax^r can not be omitted.

A more general version of Theorem 2.6.1 was given by Stens in 1980. He proved the following theorem. The case $\alpha = 0$ and $\alpha = \sigma$ are very illustrative.

Theorem 2.6.5 (Stens, [348], [349]). *Let $s \in \mathbb{N}$ and $0 \leq \alpha \leq \sigma < s$. For $f \in C[-1,1]$ the following assertions are equivalent:*

(i) *There exists a sequence $\{P_n\}$, $P_n \in \mathbb{P}_n$ such that, for $x \in [-1,1]$,*

$$| f(x) - P_n(x) | \leq C \left(\frac{(\sqrt{1 - x^2})^{\sigma - \alpha}}{n^\sigma} \right).$$

(ii) *For $\varphi(x) = \sqrt{1 - x^2}$,*

$$\sup_{|h| \leq t} \|\varphi^\alpha \Delta_h^s f\|_{C[-1,1]} \leq C t^\sigma.$$

In [150] Gopengauz analyzed the following questions. Is it possible to obtain a Timan-type estimate but improving the rate of approximation in a certain interior point? Is it possible to obtain a speed of approximating at the ends of the segment greater than the one in Timan's theorem? He considered that the rate of approximation on the whole segment is retained. Both questions were answered in the negative. For instance, he proved that an estimation of the form

$$| f(x) - p_n(x) | \leq C \omega \left(f, \frac{\psi_1(| x - a |) + \psi_2(1/n)}{n} \right),$$

is not possible for all $f \in C[-1,1]$, where $| a | < 1$ and ψ_i is an increasing function satisfying $\psi_i(t) \to 0$ as $t \to 0$ $(i = 1, 2)$. Moreover, we can not replace the expression $\Delta_n(x)$ by $o(\Delta_n(x))$.

In 1985 Yu showed some inequalities which are not possible. In particular, the following is proved.

Theorem 2.6.6 (Yu, [410]). *Let $r \in \mathbb{N} \cup \{0\}$ and $C > 0$. Then there exists a function $f \in C^r[-1,1]$ such that there exists no polynomial $P_n \in \Pi_n$ satisfying*

$$|f(x) - P_n(x)| \leq C(\sqrt{1 - x^2}/n)^r \omega_{r+3}(f^{(r)}, \sqrt{1 - x^2}/n).$$

An analogous result is stated for the case in which the quantity $\sqrt{1 - x^2}/n$ is replaced by $\sqrt{1 - x^2}/n + \epsilon_n/n^2$, ϵ_n a positive number null sequence.

In 2000 Gonska, Leviatan, Shevchuk and Wenz presented a result in the following form.

Theorem 2.6.7 (Gonska, Leviatan, Shevchuk and Wenz, [148]). *Let $k \leq r+2$ and assume that $f \in C^r[-1,1]$. Then there is a polynomial $p \in \Pi_{2[(r+k+1)/2]-1}$ for which*

$$| f(x) - p(x) | \leq C_r(\sqrt{1 - x^2})^r \omega_k(f^{(r)}, \sqrt{1 - x^2}), \quad -1 \leq x \leq 1, \qquad (2.19)$$

where C_r depends only on r. Moreover, for each $f \in C^r[-1,1]$ and $n \geq 2[(r+k+1)/2] - 1$, there is a polynomial $P_n(f) \in \mathbb{P}_n$, such that

$$| f(x) - P_n(f,x) | \leq C(r) \, (\delta_n(x))^r \, \omega_k \left(f^{(r)}, \delta_n(x) \right) \tag{2.20}$$

holds with a constant C_r depending only on r.

It should be noted that it is impossible to replace $2[(r + k + 1)/2] - 1$ by any lower figure. It has been shown by Yu [410] (see also Li [235]), that (2.20) is not valid if $k > r + 2$: assume that $k > r + 2 \geq 2$, then for each n and every constant $A > 0$, there exists a function $f = f_{r,k,n,A} \in C^r[-1,1]$ such that, for any polynomial $p_n \in \mathbb{P}_n$, there is a point $x \in [-1,1]$ for which

$$| f(x) - p_n(x) | > A(\delta_n(x))^r \omega_k(f^{(r)}, \delta_n(x))$$

holds. One also has

Theorem 2.6.8 ([148]). *Given $r \geq 0$, there exists a function $f \in C^r[-1,1]$ such that, for any algebraic polynomial p,*

$$\lim_{x \to -1} \sup \frac{| f(x) - p(x) |}{(\sqrt{1-x^2})^r \omega_{r+3}(f^{(r)}, \sqrt{1-x^2})} = \infty.$$

It is possible to construct a function which exhibits this phenomenon at both endpoints.

2.7 Characterization of some classes of functions

As we recall, if $f \in C[0, 2\pi]$, $r \in \mathbb{N}$ and $\alpha \in (0,1)$, then $f \in C^r[0, 2\pi]$ and $f^{(r)} \in \text{Lip}_\alpha$ if and only if $E_n(f)^* = \mathcal{O}(n^{-(r+\alpha)})$. The same result is not true in the non-periodic case. Some characterizations appeared in works of Timan and Dzyadyk.

Theorem 2.7.1. *Let $f \in C[-1,1]$, r a positive integer and $\alpha \in (0,1)$. The following assertions are equivalent:*

i) *$f \in C^r[-1,1]$ and for each $x \in [-1,1]$,*

$$\sup_{\{h: |h| \leq \delta, |x+h| \leq 1\}} | f^{(r)}(x) - f^{(r)}(x+h) | \leq C \, \delta,$$

where C is a positive constant which does not depend on x or δ;

ii) *For each $n \in \mathbb{N}$ there exists $P_n \in \mathbb{P}_n$ such that, for each $x \in [-1,1]$,*

$$| f(x) - P_n(x) | \leq \frac{D}{n^{r+\alpha}} \left(\sqrt{1-x^2} + \frac{1}{n} \right)^{r+\alpha},$$

where D is a positive constant which does not depend on x or δ.

In 1957 Timan presented a converse result assuming that an estimate in terms of the function $\Delta_n(x)$ is known. For the definition of modulus of continuity see (1.7).

Theorem 2.7.2 (Timan, [376]). *Let ω be a modulus of continuity and $f : [-1, 1] \to \mathbb{R}$ be a function and suppose there exists a sequence $\{P_n\}$ ($P_n \in \mathbb{P}_n$) such that*

$$| f(x) - P_n(x) | \leq \omega \left(\frac{1}{n} \left(\sqrt{1 - x^2} + \frac{|x|}{n} \right) \right), \qquad x \in [-1, 1].$$

Then

$$\omega(f, t) \leq C t \int_t^1 \frac{\omega(s)}{s^2} ds, \qquad 0 < t \leq \frac{1}{2},$$

where C is a fixed constant. Moreover, assume that $\int_0^1 (\omega(s)/s) ds < \infty$ and there exists a sequence $\{P_n\}$ ($P_n \in \mathbb{P}_n$) such that, for $x \in [-1, 1]$,

$$| f(x) - P_n(x) | \leq \frac{1}{n^r} \left(\sqrt{1 - x^2} + \frac{|x|}{n} \right)^r \omega \left(\frac{1}{n} \left(\sqrt{1 - x^2} + \frac{|x|}{n} \right) \right). \qquad (2.21)$$

Then $f \in C^r[-1, 1]$ and

$$\omega(f^{(r)}, t) \leq C \left(\int_0^t \frac{\omega(s)}{s} ds + t \int_t^1 \frac{\omega(s)}{s^2} ds \right), \qquad 0 < t \leq \frac{1}{2}.$$

Also a more general assertion can be proved:

$$\omega_r(f, t) \leq C t^r \int_t^1 \frac{\omega(s)}{s^{r+1}} ds, \qquad 0 < t \leq \frac{1}{2}.$$

Theorem 2.7.3 (Timan, [376]). *Let ω be a modulus of continuity such that*

$$\int_0^t \frac{\omega(s)}{s} ds \leq C\omega(t), \quad and \quad t \int_t^1 \frac{\omega(s)}{s^2} ds \leq C\omega(t) \qquad (2.22)$$

and let $f : [-1, 1] \to \mathbb{R}$ be a function. One has $f \in C^r[-1, 1]$ and $\omega(f^{(r)}, t) \leq C\omega(t)$ if and only if there exists a sequence $\{P_n\}$ ($P_n \in \mathbb{P}_n$) satisfying (2.21).

In order to obtain the converse result, different variants of the Bernstein inequality are needed. That is we should estimate the derivatives of an algebraic polynomial in terms of the polynomial.

Theorem 2.7.4. *Assume that $r, n \in \mathbb{N}$ and let $\| \cdot \|$ denote the uniform norm on $[-1, 1]$.*

(i) (Markov, [248]) *If $P_n \in \mathbb{P}_n$, then*

$$\| P_n^{(r)} \| \leq n^{2r} \| P_n \|.$$

(ii) (Bernstein, [27]) *If $P_n \in \mathbb{P}_n$ and $x \in [-1,1]$, then*

$$\left| \sqrt{1 - x^2}\, P_n'(x) \right| \le n \, \| P_n \|.$$

(iii) *There exists a constant C_r such that (see [379], p. 227), if $P_n \in \mathbb{P}_n$ and $x \in [-1,1]$; then*

$$\left| (\Delta_n(x))^r P_n^{(r)}(x) \right| \le C_r \, \| P_n \|.$$

(iv) (Potapov, [288]) *If $P_n \in \mathbb{P}_n$ and $x \in [-1,1]$, then*

$$\left\| (\sqrt{1 - x^2})^r P_n^{(r)}(x) \right\|_p \le C n^r \, \| P_n \|.$$

(v) *If $p \ge 0$, $q \ge 0$ and $p + q = l$, then there exists a constant C_l such that, if $P_n \in \mathbb{P}_n$ and $x \in [-1,1]$, then*

$$\left| (\Delta_n(x))^{q/2} P_n^{(l)}(x) \right| \le C_l \, n^{l+p} \, \| P_n \|. \tag{2.23}$$

Theorem 2.7.5. *Fix positive constants L and ρ.*

(i) (Dzyadyk 1956, [107]) *If for $x \in [-1,1]$ a polynomial $P_n \in \mathbb{P}_n$ satisfies the inequality*

$$| P_n(x) | \le L \left[(\sqrt{1 - x^2})^\rho + \frac{1}{n^\rho} \right],$$

then there exists a constant C (which depends only on ρ) such that, for $x \in (-1,1)$ one has

$$| P_n'(x) | \le C n L \min \left\{ (\sqrt{1 - x^2})^{\rho-1}, \frac{1}{n^{\rho-1}} \right\}, \qquad if \quad \rho \le 1$$

and

$$| P_n'(x) | \le C n L \left[(\sqrt{1 - x^2})^{\rho-1} + \frac{1}{n^{\rho-1}} \right], \qquad if \quad \rho \ge 1.$$

(ii) (Potapov 1960, [288]) *If $\rho, \gamma \in \mathbb{R}$, there exists a constant C such that, if for $x \in [-1,1]$ a polynomial $P_n \in \mathbb{P}_n$ satisfies the inequality*

$$| P_n(x) | \le L (n+1)^{\gamma+\rho} (\Delta_{n+1}(x))^\rho,$$

then for $x \in (-1,1)$ one has

$$| P_n'(x) | \le C L (n+1)^{\gamma+\rho} (\Delta_{n+1}(x))^{\rho-1}.$$

In (2.21) only integer values of r are involved. The characterization of Lipschitz functions was done by Dzyadyk.

Theorem 2.7.6 (Dzyadyk, [107]). *Assume that* $r \in \mathbb{N}_0$, $0 < \alpha < 1$ *and* C *is a positive constant. For a function* $f : [-1, 1] \to \mathbb{R}$ *the following assertions are equivalent*:

(i) *For each* $n \in \mathbb{N}$, *there exists* $P_n \in \mathbb{P}_n$ *such that*

$$| f(x) - P_n(x) | \le \frac{C}{n^{r+\alpha}} \left((\sqrt{1-x^2})^{r+\alpha} + \frac{1}{n^{r+\alpha}} \right). \tag{2.24}$$

(ii) $f \in C^r[-1, 1]$ *and* $f^{(r)} \in \mathrm{Lip}_\alpha[-1, 1]$.

Proof. (ii) $\Longrightarrow$ (i) follows from Timan's Theorem 2.4.1.

(i) $\Longrightarrow$ (ii). Let us first consider the case $r = 0$. Fix $h < 0$ and $x \in [0, 1)$ such that $x + h \in (0, 1]$. Let us write

$$U_{2^{i+1}}(x) = P_{2^{i+1}}(x) - P_{2^i}(x), \qquad i = 0, 1, \ldots.$$

Notice that $f(x) = P_1(x) + \sum_{i=1}^{\infty} U_{2^i}(x)$.

For any $k \in \mathbb{N}$ one has

$$| f(x+h) - f(x) | \le | P_1(x+h) - P_1(x) | + \sum_{i=0}^{k-1} | U_{2^{i+1}}(x+h) - U_{2^{i+1}}(x) |$$

$$+ \sum_{i=k}^{\infty} | U_{2^{i+1}}(x+h) | + \sum_{i=k}^{\infty} | U_{2^{i+1}}(x) | .$$

From (i) we obtain

$$| U_{2^{i+1}}(x) | \le | P_{2^{i+1}}(x) - f(x) | + | f(x) - P_{2^i}(x) |$$

$$\le \frac{2^{1+\alpha}C_1}{2^{i\alpha}} \left[(\sqrt{1-x^2})^\alpha + \frac{1}{2^{(i+1)\alpha}} \right].$$

Therefore

$$\sum_{i=k}^{\infty} | U_{2^{i+1}}(x) | \le 2C_1 (\sqrt{1-x^2})^\alpha \frac{2^{2\alpha}}{2^\alpha - 1} \frac{1}{2^{k\alpha}} + 2C_1 \frac{4^\alpha}{4^\alpha - 1} \frac{1}{2^{k\alpha}}$$

$$= \frac{2^{1+3\alpha}C_1}{2^\alpha - 1} \frac{(\sqrt{1-x^2})^\alpha}{2^{(k+1)\alpha}} + \frac{4^{2\alpha+1/2}}{4^\alpha - 1} \frac{C_1}{2^{2(k+1)\alpha}}.$$

Now we consider two cases.

Case 1. Assume first that $x \ge 0$ and $x \in [1 - 2h, 1]$. Fix k such that

$$2^k \le \frac{1}{\sqrt{h}} < 2^{k+1}.$$

From the arguments given above we know that, if $\xi \in [1 - 2h, 1]$, then

$$\sum_{i=k}^{\infty} | U_{2^{i+1}}(\xi) | \leq \frac{2^{1+3\alpha} C_1}{2^\alpha - 1} \frac{(\sqrt{1 - (1 - 2h)^2})^\alpha}{2^{(k+1)\alpha}} + \frac{4^{2\alpha+1/2}}{4^\alpha - 1} \frac{C_1}{2^{2(k+1)\alpha}}$$

$$\leq \frac{2^{1+4\alpha} C_1}{2^\alpha - 1} h^{\alpha/2 + \alpha/2} + \frac{4^{2\alpha+1/2}}{4^\alpha - 1} h^\alpha < \frac{64 C_1}{2^\alpha - 1} h^\alpha.$$

On the other hand, taking into account Theorem 2.7.5 we obtain

$$\sum_{i=0}^{k-1} | U_{2^{i+1}}(x + h) - U_{2^{i+1}}(x) | \leq h \sum_{i=0}^{k-1} | U'_{2^{i+1}}(x + h\theta_i) |$$

$$\leq C_2 2^{1+\alpha} h \sum_{i=0}^{k-1} \frac{1}{2^{i\alpha}} 2^{(i+1)(2-\alpha)} = C_3 h \sum_{i=0}^{k-1} 2^{i(1-\alpha)} \leq C_4 h \, 2^{k(1-\alpha)}.$$

Thus, for $x \in [1 - 2h, 1 - h]$ we have proved that there exists a constant K such that

$$| f(x + h) - f(x) | \leq K h^\alpha.$$

Case 2. Assume that $x \geq 0$ and $x \in [0, 1 - 2h]$. Fix k such that

$$2^k \leq \frac{\sqrt{1 - x^2}}{h} < 2^{k+1}.$$

Notice that

$$\frac{1}{2^{k+1}} < \frac{h}{\sqrt{1 - x^2}} \leq \frac{h}{\sqrt{1 - (1 - 2h)^2}} = \frac{h}{\sqrt{4h(1 - h)}} \leq \sqrt{\frac{h}{2}}.$$

In this case, if $\xi \in [x, 1 - h]$, then

$$\sum_{i=k}^{\infty} | U_{2^{i+1}}(\xi) | \leq \frac{2^{1+3\alpha} C_1}{2^\alpha - 1} (\sqrt{1 - \xi^2})^\alpha \frac{h}{(\sqrt{1 - \xi^2})^\alpha} + \frac{4^{2\alpha+1/2} C_1}{4^\alpha - 1} \left(\sqrt{\frac{h}{2}}\right)^{2\alpha} \leq C_5 h^\alpha.$$

For the estimate of the sum for $0 \leq i \leq k$, notice that for $x \in [0, 1 - 2h]$ and $0 < \theta < 1$

$$\frac{1 - x^2}{2} \leq (1 - (x + h)^2 \leq (1 - (x + h\theta)^2$$

and

$$2h \leq 4h(1 - h) = 1 - (1 - 2h)^2 \leq 1 - x^2.$$

Hence

$$\sum_{i=0}^{k-1} |\, U_{2^{i+1}}(x+h) - U_{2^{i+1}}(x)\,| \le h \sum_{i=0}^{k-1} |\, U'_{2^{i+1}}(x+h\theta_i)\,|$$

$$\le C_6\, h \sum_{i=0}^{k-1} \frac{1}{(\sqrt{1-(x+h\theta_i)^2})^{1-\alpha}} \frac{2^{i+1}}{2^{i\alpha}} \le C_7\, h \sum_{i=0}^{k-1} 2^{i(1-\alpha)} \left(\sqrt{\frac{1-x^2}{2}}\right)^{\alpha-1}$$

$$\le C_8\, h \frac{2^{k(1-\alpha)}}{(\sqrt{1-x^2})^{1-\alpha}} \le C_9\, h \frac{1}{(\sqrt{1-x^2})^{1-\alpha}} \frac{(\sqrt{1-x^2})^{1-\alpha}}{h^{1-\alpha}} = C_9\, h^\alpha.$$

The theorem is proved for the case $r = 0$.

For $r > 0$ we differentiate the representation of f to obtain $f^{(r)}(x) = P_1^{(r)}(x) + \sum_{i=1}^{\infty} U_{2^i}^{(r)}(x)$. Then use Theorem 2.7.5 to obtain the inequality

$$|\, U_{2^{i+1}}^{(r)}(x)\,| \le \frac{C}{2^{i\alpha}} \left[(\sqrt{1-x^2})^\alpha + \frac{1}{2^{(i+1)\alpha}}\right].$$

Then we can use arguments similar to ones for the case $r = 0$.   □

With respect to the Zygmund class, Dzyadyk proved the following:

Theorem 2.7.7 (Dzyadyik, [107]). *Assume that $r \in \mathbb{N}_0$, $0 < \alpha < 1$. If for a function $f : [-1,1] \to \mathbb{R}$ there exists a sequence $\{P_n\}$ $(P_n \in \mathbb{P}_n)$ such that*

$$|\, f(x) - P_n(x)\,| \le \frac{C}{n^{r+1}} \left((\sqrt{1-x^2})^{r+1} + \frac{1}{n^{r+1}}\right)$$

where C does not depend on n, then $f \in C^r[-1,1]$ and $f^{(r)} \in Z[-1,1]$.

In 1960 Potapov obtained a characterization related with the first modulus.

Theorem 2.7.8 (Potapov, [289]). *For $f \in C[-1,1]$ one has $E_n(f) = \mathcal{O}(n^{-\alpha})$ if and only if*

$$|\, f(\cos(\theta + t)) - f(\cos\theta)\,| \le C\, |\, t\,|^\alpha,$$

where C is a positive constant which does not depend on θ or t.

This result clearly shows that if for $f \in C[-1,1]$ one has $E_n(f) = \mathcal{O}(n^{-\alpha})$, then inside the segment f satisfies a Lipschitz condition of order α and in the end of the segment a Lipschitz condition of order $\alpha/2$.

The results of Timan and Dzyadyk seems to be of a point-wise nature. Some authors tried to put them as estimates in norm, but they used varying weights. For instance, Scherer-Wagner [330] defined the weighted best approximation by

$$E_n^{(r,\alpha)}(f) = \inf_{p\in\mathbb{P}_n} \left\| \frac{f(x) - p(x)}{(n\Delta_n(x))^{r+\alpha}} \right\|_C \tag{2.25}$$

and proved that (see Golitschek [143] for similar results concerning $L_p(-1,1)$)

$$E_n^{(r,\alpha)}(f) = \mathcal{O}(n^{-(r+\alpha)}) \Leftrightarrow f^{(r)} \in C[-1,1] \quad \text{and} \quad \omega_1(f^{(r)}, 1/n) \leq Cn^{-\alpha}.$$

Teliakovskii used Theorem 2.6.1 to obtain a characterization theorem using the function $\delta_n(x)$.

Theorem 2.7.9 (Teliakovskii, [371]). *Let r be a non-negative integer and $f:[-1,1]\to \mathbb{R}$ a function.*

(i) *Let w be a modulus of continuity satisfying (2.22). One has $f \in C^r[-1,1]$ and $\omega(f^{(r)}, t) \leq C(f)w(t)$ if and only if, for each $n > r$ there exists $P_n \in \mathbb{P}_n$ such that*

$$| f(x) - P_n(f,x) | \leq C(f) \, (\delta_n(x))^r \, \omega \, (\delta_n(x)) .$$

(ii) *If $\alpha \in (0,1)$, one has $f \in C^r[-1,1]$ and $f^{(r)} \in \mathrm{Lip}_\alpha[1,1]$ if and only if, for each $n > r$ there exists $P_n \in \mathbb{P}_n$ such that*

$$| f(x) - P_n(f,x) | \leq C(f) \, (\delta_n(x))^{r+\alpha} .$$

Other classes can be characterized. Let us consider functions ψ satisfying the following condition:

$$\int_0^t \frac{\psi(u)}{u} du + t^k \int_t^1 \frac{\psi(u)}{u^{k+1}} du \leq C\psi(u). \tag{2.26}$$

This kind of function has been used in the works of Stechkin [347], Lozinskii [241] and Bari and Stechkin [15] to present results for the approximation of periodic functions.

Let us write

$$W^r H_k[\psi] = \{ f \; : \; \omega(f^{(r)}, t) \leq C(f)\psi(t) \}.$$

Theorem 2.7.10. *Fix ψ such that (2.26) holds. If for a function f and every $n \in N$ there exists a polynomial P_n such that (2.15) is satisfied, then $f \in W^r H_k[\psi]$.*

Thus, if $E_n(f) = \mathcal{O}(n^{-2r}\varphi(n^{-2}))$, then $f \in W^r H_k[\psi]$.

Notice that for $\varphi(t) = t^\alpha$ one has Dzyadyk's theorem. As Shevchuk showed, the converse of the last result is not true.

Theorem 2.7.11 (Shevchuk, [338]). *Suppose that ψ does not satisfy (2.26).*

- *There exists a function f for which $E_n(f)=\mathcal{O}(n^{-2r}\psi(n^2))$ and $f \notin W^r H_k[\psi]$.*
- *There exists a function $f \notin W^r H_k[\varphi]$ and a sequence $\{P_n\}$ of polynomials such that the Timan estimate holds.*

For $r = 1$, a significantly stronger result (which, in particular, implies Theorem 2.7.11 for $k = 1$, $r = 0$) was obtained earlier by Dolzhenko and Sevastyanov [103].

Theorem 2.7.12 (Shevchuk, [338]). *For any function $\varphi \in \Phi^k$, there is a function $f \in W^r H_k^\varphi$ such that*

(i) *For all $n \in N$, $E_n(f) \leq n^{-2r} \varphi(n^{-2})$,*

(ii) *$\omega_k(f^{(r)}, t) \geq c\varphi(t)$, $t \in [0, 1/k]$, $c = c(r, k) > 0$.*

Theorem 2.7.13 ([338]). *Let a_n be an increasing sequence of natural numbers, such that $\sum_{n=1}^{\infty} (na_n)^{-1} = \infty$. There exists a function $f \in C[0, 1]$ for which $f \notin C^r[0, 1]$ and $E_n(f) = \mathcal{O}(n^{-2r}/a_n)$ $(r \geq 2)$.*

Theorem 2.7.13 was proved by another method in a paper of Xie [412]. The theorem was stated as a conjecture in the work [157] of Hasson. For the case $\varphi(t) = t^\alpha$, $0 < \alpha < k$, $\alpha \notin N$, Theorem 2.7.12 follows from results of Bernstein on the approximation of the function $(1 - x)^\alpha$ on $[-1, 1]$. A proof of Theorem 2.7.12 for the indicated case can also be found in [157]. For $\varphi(t) = t^\alpha$, $0 < \alpha < k$, $\alpha \in N$ this theorem follows from results of Ibragimov. In connection with Theorem 2.7.12 we note the following example of Brudnyi [38]. The continuous function $f_{a,b} : [0, 1] \to \mathbb{R}$, defined on $(0, 1]$ by the formula $f_{a,b}(x) = x^a \sin x^{-b}$, $a, b > 0$, has for $k > a/(1 + b)$ the modulus of continuity $\omega(f_{a,b})(t) = t^{a/(1+b)}$, whereas $E_n(f) \sim n^{-2a/(2b+1)}$. Theorems 2.7.11 and 2.7.12 show, in particular, that the assertion of the inverse theorem cannot be sharpened for any of the classes $W^r H_k[\varphi]$ if the rate of approximation is characterized not by the quantities $\rho_n(x)$ but rather by n^{-2}.

2.8   Simultaneous approximation

First, let us recall some facts related with trigonometric approximation. The approximation of the derivatives of a function by the derivatives of the polynomial which approximate the function was considered by Freud [122]. He proved that, for any polynomial T_n,

$$\|f^{(r)} - T_n^{(r)}\| \leq C_r \{n^r \|f - T_n\| + E_n^*(f^{(r)})\}, \tag{2.27}$$

where C_r is a constant which depends only on r. A related inequality was given by Czipszer and Freud in [78].

Theorem 2.8.1. *Fix $k \in \mathbb{N}$.*

(i) *There exists a constant K such that, if $f \in C^k[0, 2\pi]$ and $T_n \in \mathbb{T}_n$, then*

$$\|f^{(k)} - T_n^{(k)}\| \leq K \log (1 + \min\{k, n\}) \{n^r \|f - T_n\| + E_n^*(f^{(r)})\}.$$

Moreover, if $\|f - T_n\| \leq CE_n^*(f)$, then

$$\|f^{(k)} - T_n^{(k)}\| \leq KCE_n^*(f^{(r)}). \tag{2.28}$$

(ii) *There exists a constant K such that, if $f, f^{(k)} \in L_1[0, 2\pi]$ and $T_n \in \mathbb{T}_n$ satisfies $\|f - T_n\|_1 \leq CE_n^*(f)_1$, then*

$$\|f^{(k)} - T_n^{(k)}\|_1 \leq CK \log\left(1 + \min\{k, n\}\right) E_n^*(f^{(r)})_1.$$

(iii) *For each $p \in (1, \infty)$, there exists a constant B_p such that, if $f, f^{(k)} \in L_p[0, 2\pi]$ and $T_n \in \mathbb{T}_n$ satisfies $\|f - T_n\|_p \leq CE_n^*(f)_p$, then*

$$\|f^{(k)} - T_n^{(k)}\|_p \leq CB_p E_n^*(f^{(r)})_p.$$

The inequality (2.27) was improved by Garkavi [133]. Set

$$C_{n,r}(f) = \inf_{T_n \in \mathbb{T}_n} \max_{1 \leq k \leq r} \frac{\|f^{(k)} - T_n^{(k)}\|}{E_n^*(f^{(k)})} \quad \text{and} \quad C_{n,r} = \sup_{f \in W^r(1,[0,2\pi])} C_{n,r}(f).$$

Garkavi proved that

$$C_{n,r} = \frac{4}{\pi^2}\left(\ln(p+1)\right) + \mathcal{O}(\ln \ln \ln p)),$$

where $p = \min\{n, r\}$ and

$$\|f^{(r)} - T_n^{(r)}\| \leq n^r \|f - T_n\| + \left(1 + \frac{\pi}{2}\right) C_{n,r} E_n(f^{(r)}).$$

One of the first results on simultaneous approximation is due to Gelfond in 1955.

Theorem 2.8.2 (Gelfond, [141]). *If $f \in C^m[a, b]$, for $n \geq n_0$, there exists $P_n \in \mathbb{P}_n$ such that*

$$\|f^{(k)} - P_n^{(k)}\| \leq C\frac{1}{n^{m-k}} \omega\left(f^{(m)}, \frac{1}{n}\right), \qquad (0 \leq k \leq m).$$

Theorem 2.8.3 (Feinerman and Newman, [117]). *There exists a constant K such that, if $f \in C^1[a, b]$, then*

$$E_n(f) \leq \frac{K}{n} E_{n-1}(f') \qquad n \geq 1. \tag{2.29}$$

Hasson found estimates in norms in the spirit of Garkavi's results.

Proposition 2.8.4 (Hasson, [156]). *There exists a constant M with the following property: Let $f \in C[a, b]$ be such that, for some λ, $E_n(f) \leq \lambda/n$, $n \geq 1$, $E_n(f) \leq \lambda$. Then, if P_n is the polynomial of best approximation to f, one has*

$$\|P_n'\| \leq M\lambda n, \qquad n \geq 1.$$

Proof. Fix k such that $2^k \leq n < 2^k$. By differentiating the identity

$$P = P_n - P_{2^k} + \sum_{i=1}^{k}(P_{2^i} - P_{2^{i-1}}) + (P_1 - P_0) + P_0$$

and applying the Markov inequality we obtain

$$\|P'\| \leq K\left(n^2\|P_n - P_{2^k}\| + \sum_{i=1}^{k}2^{2i}\|P_{2^i} - P_{2^{i-1}}\| + (P_1 - P_0)\right)$$

$$\leq K\left(2n^2 E_{2^k}(f) + \sum_{i=1}^{k}2^{2i+1}E_{2^{i-1}}(f) + 2E_0(f))\right)$$

$$\leq K\left(2\frac{2^{2(k+1)}}{2^k}\lambda + \sum_{i=1}^{k}2^{2i+1}\frac{\lambda}{2^{i-1}} + 2\lambda)\right)$$

$$\leq K\lambda\left(82^k + 4\sum_{i=1}^{k}2^i + 2)\right) \leq M\lambda n. \qquad \square$$

Theorem 2.8.5 (Hasson, [156]). *Let k and r be integers. For $f \in C^r[a,b]$, let $P_n(f) \in \mathbb{P}_n$ be the polynomial of best approximation for f. There exist constants M, S and T depending on r such that*

$$\|f^{(k)} - P_n^{(k)}(f)\| \leq M n^k E_{n-k}(f^{(k)}), \qquad 0 \leq k \leq r, \quad n \geq k,$$
$$\|P_n^{(k)}(f)\| \leq \|f^{(k)}\| + M n^k E_{n-k}(f^{(k)}), \quad 0 \leq k \leq r, \quad n \geq k, \qquad (2.30)$$

and

$$\|f^{(k)} - P_n^{(k)}(f)\| \leq SE_{n-2k}(f^{(2k)}) \leq TE_{n-r}(f)\frac{1}{n^{r-2k}}E_{n-r}(f^{(r)}),$$

for $0 \leq k \leq m/2$ and $n \geq m$.

Moreover (Roulier, [314])

$$\|f^{(k)} - P_n^{(k)}(f)\| \leq M\frac{1}{n^{r-2k}}\omega\left(f^{(r)}\frac{1}{n}\right), \qquad n > r.$$

Proof. It is clear that (2.30) holds for $k = 0$. Assume that (2.30) holds for r. By induction, for $f \in C^{r+1}[0,1]$ one has

$$\|f^{(k+1)} - Q_{n-1}^{(k)}\| \leq M n^k E_{n-1-k}(f^{(k+1)}), \qquad 0 \leq k \leq r, \quad n \geq k,$$

where Q_{n-1} is the polynomial of best approximation to f'. If we set

$$g(x) = f(x) - f(a) - \int_a^x Q_{n-1}(t)dt, \qquad x \in [a,b],$$

then, for $a \leq x < y \leq b$,

$$| g(x) - g(y) | \leq \int_x^y | f'(t) - Q_{n-1}(t) | \, dt \leq E_{n-1}(f') \, | x - y | .$$

Let R_n be the polynomials of best approximation to g. From the direct estimate and Proposition 2.8.4 we know that

$$\| R_n' \| \leq K_1 n \, E_{n-1}(f'), \qquad n \geq 1,$$

and, using Markov inequality and (2.29), we obtain

$$\| R_n^{(k)} \| \leq K_k \, n \, n^{2(k-1)} \, E_{n-1}(f')$$
$$\leq K_k^* \frac{n^{2k-1}}{(n-1)(n-2) \cdots (n-(k-1))} E_{n-k}(f^{(k)})$$
$$\leq K_k' \, n^k \, E_{n-k}(f^{(k)}), \qquad 0 \leq k \leq r+1, \quad n \geq k.$$

Therefore

$$\| f^{(k)} - Q_{n-1}^{(k-1)} - R_n^{(k)} \| \leq K_k' \, n^k E_{n-k}(f^{(k)}) + M_r n^{k-1} \, E_{n-k}(f^{(k)})$$
$$\leq M_{r+1} \, n^k E_{n-k}(f^{(k)}), \qquad 0 \leq k \leq r+1, \quad n \geq k.$$

The result follows because $-f(a) + \int_a^x Q_{n-1}(t) dt + R_n(x)$ is the polynomial of best approximation to f.

The last assertion follows from Jackson's theorem. In fact

$$E_{n-r}(f^{(r)}) \leq C \omega \left(f^{(r)} \frac{1}{n-r} \right) \leq C \left(1 + \frac{1}{n-r} \right) \omega \left(f^{(r)} \frac{1}{n} \right). \qquad \square$$

Theorem 2.8.6 ([156]). *Let $a < c < d < b$ and let m and k be integers with $0 \leq k \leq m$. There exists a constant C, which depends on m, c and d such that, if P_n is the polynomial of best approximation to $f \in C^m[a,b]$, then*

$$\| f^{(k)} - P_n^{(k)} \|_{[c,d]} \leq C E_{n-k}(f^{(k)}), \qquad n \geq k.$$

Theorem 2.8.7 ([156]). *Let k and r be integers, $k > r \geq 0$. Fix $f \in C^r[a,b]$ and, for each $n \in \mathbb{N}$, let $P_n(f) \in \mathbb{P}_n$ be the polynomial of best approximation for f. If f is not a polynomial, there exist constants $M(f,k)$, such that*

$$\| P_n^{(k)}(f) \| \leq M(f,r) \, n^{2k-r} \, \omega \left(f^{(r)}, \frac{1}{n} \right) \qquad n \geq 1.$$

Proof. The proof of this theorem is based on an extension of f. Fix two reals c and d ($c < a$ and $b > d$) and assume that f has been extended to a function $F \in C^r[c,d]$ in such a way that $\omega(F^{(r)}, h) \leq C \omega(f^{(r)}, h)$.

Fix a sequence $\{Q_n\}$ of polynomials such that

$$\|Q_n^{(k)} - F^{(k)}\|_{[c,d]} \le \frac{K}{n^{r-k}} \omega\left(F^{(r)}, \frac{1}{n}\right) \le \frac{CK}{n^{r-k}} \omega\left(f^{(r)}, \frac{1}{n}\right),$$

for $k \le r$ and $n \ge k + 1$. One has $\|Q_n^{(k)}\|_{[c,d]} \le C_k$, for $0 \le k \le r$ and (by Bernstein's inequality) $\|Q_n^{(k)}\|_{[c,d]} \le K_k n^{k-r}$, for $k > r$. Since, for $k \ge 0$,

$$\|P_n^{(k)}\|_{[a,b]} \le \|P_n^{(k)} - Q_n^{(k)}\|_{[a,b]} + \|Q_n^{(k)}\|_{[a,b]}$$

and

$$\|P_n^{(k)} - Q_n^{(k)}\|_{[a,b]} \le S_k\, n^{2k}[\|P_n - Q_n\|_{[a,b]} \le N_k n^{2k}\left(E_n(f) + \frac{Kl}{n^r}\omega(f^{(r)}, \frac{1}{n})\right),$$

Jackson's inequality yields

$$\|P_n^{(k)}\|_{[a,b]} \le C_5\, n^{2k-r} \omega\left(f^{(r)}, \frac{1}{n}\right) + \max\left(K_k, K_k n^{k-r}\right).$$

Thus the proof finishes by proving that the second term can be estimated with the first one.   □

Trigub was one of the first in considering a point-wise estimate for simultaneous approximation by algebraic polynomials. He also noticed that we can use the second-order modulus, instead of the first one, and provided some inequalities for the derivatives of the polynomials. In 1968 Malozemov [246] proved that the constant in the corresponding estimates of Gelfond and Trigub do not depend on the functions. We present the assertion as it appeared in a paper of Malosemov [245].

Theorem 2.8.8 (Trigub, [388]). *If $f \in C^r[-1, 1]$, then for each $n \in \mathbb{N}$ there exists a polynomial $P_n \in \mathbb{P}_n$ such that, for all $x \in [-1, 1]$ and $k = 0, 1, \ldots, r$,*

$$| f^{(k)}(x) - P_n^{(k)}(x) | \le C_r\, (\Delta_n(x))^{r-k}\, w\left(f^{(r)}, \Delta_n(x)\right) \tag{2.31}$$

where C_r does not depend upon n or f.

Is the last result a consequence of the particular polynomials used in the approximation? In 1966 Teliakovskii showed that, for a differentiable function f, the derivatives of any sequence of the polynomials which approximate f with the rate given in Timan's theorem, approximate f' with a similar rate.

We need an estimate for the derivatives of polynomials.

Proposition 2.8.9. *There exists a constant R with the following property: let $a \ge 0$ be a real number, $r \ge 1$ an integer and ω a modulus of continuity. If a polynomial P_n satisfies the inequality*

$$| P_n(x) | \le (\Delta_n(x))^r\, \omega(\Delta_n(x)) + a, \qquad x \in [-1, 1],$$

then

$$| P_n'(x) | \le R\left((\Delta_n(x))^{r-1}\, \omega(\Delta_n(x)) + a(\Delta_n(x))^{-1}\right), \qquad x \in [-1, 1].$$

The last result was proved by Lebed [225] in the case $a = 0$. Other proofs were given in [107] and [379] (p. 219–226). According to Teliakovskii [371], for $a > 0$ the proof can be obtained with arguments similar to the one used in [379].

Theorem 2.8.10 (Teliakovskii, [371]). *Assume* $r \in \mathbb{N}_0$ *and* $f \in C^r[-1,1]$. *If* $\{P_n(f,x)\}$ *is a sequence of polynomials satisfying* (2.9), *then for* $k = 1, \ldots, r$,

$$| f^{(k)}(x) - P_n^{(k)}(f,x) | \leq C_{r,k} \, (\Delta_n(x))^{r-k} \, \omega\left(f^{(r)}, \Delta_n(x)\right),$$

where the constant $C_{r,k}$ *does not depend upon* f *or* n.

Proof. We only present the main ideas of the proof.

Let $\{P_n\}$ be a sequence of polynomials for which the Timan estimate (2.9) holds. For $s \in \mathbb{N}_0$, write $n_s = 2^s n$, $p_0 = P_n$ and $p_s = P_{n_s}$. From the identity

$$f(x) - p_0(x) = \sum_{s=1}^{\infty} [p_s(x) - p_{s-1}(x)]$$

we obtain

$$| f^{(k)}(x) - p_0^{(k)}(x) | = \left| \sum_{s=1}^{\infty} [p_s^{(k)}(x) - p_{s-1}^{(k)}(x)] \right|$$

$$\leq (RA + R) \sum_{s=1}^{\infty} \left(\frac{\sqrt{1-x^2}}{n_s} + \frac{1}{n_s} \right)^{r-k} \omega\left(f^{(r)}, \frac{\sqrt{1-x^2}}{n_s} + \frac{1}{n_s} \right)$$

$$\leq (RA + R) \sum_{s=1}^{\infty} \left(\frac{\sqrt{1-x^2}}{2^s n} + \frac{1}{4^s n^2} \right)^{r-k} \omega\left(f^{(r)}, \frac{\sqrt{1-x^2}}{2^s n} + \frac{1}{4^s n^2} \right).$$

If $k < r$, then

$$| f^{(k)}(x) - P_n^{(k)}(x) | \leq C \, (\Delta_n(x))^{r-k} \, \omega\left(f^{(r)}, \Delta_n(x)\right) \sum_{s=1}^{\infty} \frac{1}{2^{s(r-k)}}.$$

The theorem is proved for $k < r$.

For the case $k = r$, it is sufficient to consider the case $r = 1$.

Assume $r = 1$ and fix a point x_0 and set $h = \Delta_n(x_0)$. There exists a function $F_h(f) \in C^1[-1,1]$ such that

$$| f(x) - F_h(f,x) | \leq \frac{1}{2} h \omega(f', h), \tag{2.32}$$

$$| f'(x) - F_h'(f,x) | \leq \omega(f', h) \tag{2.33}$$

and

$$\omega(F_h', t) \leq \begin{cases} \delta \, \omega(f', h)/h, & \text{if} \quad \delta \leq h, \\ 3\omega(f', h), & \text{if} \quad h < \delta. \end{cases} \tag{2.34}$$

Then
$$\begin{aligned} \mid f'(x) - p_n'(x) \mid &\leq \mid f'(x) - F_h'(x) \mid + \mid F_h'(x) - p_n'(x) \mid \\ &\leq \omega(f', h) + \mid F_h'(x) - p_n'(x) \mid . \end{aligned} \tag{2.35}$$

There exist polynomials Q_m such that

$$\mid F_h'(x) - Q_m(x) \mid \leq C\,\Delta_m(x)\,\omega(F_h', \Delta_m(x)). \tag{2.36}$$

Now, we use the representation

$$F_h(x) - p_n(x) = \sum_{s=1}^{\infty} [Q_{n_s}(x) - Q_{n_{s-1}}(x)] + Q_n(x) - p_n(x). \tag{2.37}$$

In this case we have

$$\mid Q_{n_s}(x) - Q_{n_{s-1}}(x) \mid \leq C\,\Delta_{n_s}(x)\,\omega\left(F_h', \Delta_{n_s}(x)\right).$$

From Proposition 2.8.9 (with $a = 0$) we obtain

$$\mid Q_{n_s}'(x) - Q_{n_{s-1}}'(x) \mid \leq C_1\,\omega\left(F_h', \Delta_{n_s}(x)\right).$$

On the other hand, we can use (2.36), (2.32), the hypothesis (2.9) and (2.34) to estimate the difference $Q_n - p_n$. In fact

$$\begin{aligned} \mid Q_n(x) - p_n(x) \mid &\leq \mid Q_n(x) - F_h(x) \mid + \mid F_h(x) - f(x) \mid + \mid f(x) - p_n(x) \mid \\ &\leq C\,\Delta_n(x)\,\omega(F_h', \Delta_n(x)) + \frac{1}{2} h\omega(f', h) + A\Delta_n(x)\omega(f', \Delta_n(x)) \\ &\leq C_2\,\Delta_n(x)\,\omega(f', \Delta_n(x)) + \frac{1}{2} h\omega(f', h). \end{aligned}$$

From the last estimate and Proposition 2.8.9 (with $a = h\omega(f', h)/2$) we obtain

$$\mid Q_n'(x) - p_n'(x) \mid \leq C_3\,\omega(f', \Delta_n(x)) + C_4(\Delta_n(x))^{-1} h\omega(f', h).$$

Therefore the series in (2.37) converges uniformly and we can differentiate term by term. That is

$$\mid F_h'(x) - p_n'(x) \mid \leq C_1 \left(\sum_{s=1}^{\infty} \omega(F_h', \Delta_{n_s}(x)) + \omega(f', \Delta_n(x)) + (\Delta_n(x))^{-1} h\omega(f', h) \right).$$

Finally, for $x = x_0$ and $h = \Delta_n(x_0)$, from the last inequality and (2.35) one has

$$\begin{aligned} \mid f'(x_0) - p_n'(x_0) \mid &\leq C \left(\sum_{s=1}^{\infty} \frac{\Delta_{n_s}(x_0)\omega(f', h)}{h} + \omega(f', \Delta_n(x_0)) + \frac{h\omega(f', h)}{\Delta_n(x_0)} \right) \\ &\leq C \left(\omega(f', \Delta_n(x_0)) + \frac{\omega(f', h)}{h} \sum_{s=1}^{\infty} \left(\frac{\sqrt{1 - x_0^2}}{2^s n} + \frac{1}{4^s n^2} \right) \right) \\ &\leq C\omega(f', \Delta_n(x_0)). \end{aligned}$$

$\square$

In particular, from Timan's theorem Teliakovskii derived a new proof of the
Trigub result presented above. On the other hand, Theorem 2.8.10 can be obtained
from Theorem 2.8.8 and Proposition 2.8.9.

An analogue of Theorem 2.8.10, with $\delta_n(x)$ instead of $\Delta_n(x)$ is due to Gopengauz. He constructed linear polynomial operators $L_{n,r} : C^r[-1,1] \to \mathbb{P}_n$ for each
fixed $r \geq 0$, such that the following theorem holds:

Theorem 2.8.11 (Gopengauz, [150]). *For each $r \geq 0$ there exists a sequence of
linear operators $L_{n,r} : C^r[-1,1] \to \mathbb{P}_n$ ($n \geq 4r + 5$) such that, for $f \in C^r[-1,1]$
for $0 \leq k \leq r$,*

$$\mid f^{(k)}(x) - L_{n,r}^{(k)}(x) \mid \leq C_r \left(\delta_n(x)\right)^{r-k} \omega\left(f^{(r)}, \delta_n(x)\right), \tag{2.38}$$

where the constant C_r does not depend on f, n and x.

In 1978 Vértesi noticed that, under additional assumptions, one can replace
$\Delta_n(x)$ by $\delta_n(x)$.

Theorem 2.8.12 (Vértesi, [399]). *Assume $r \in \mathbb{N}_0$ and $f \in C^r[-1,1]$. If $\{P_n(f,x)\}$
is a sequence of polynomials satisfying (2.9) and*

$$P_n^{(k)}(f,\pm 1) = f^{(k)}(\pm 1), \qquad (k = 0,1,\ldots,r),$$

then for $k = 0,1,\ldots,r$,

$$\mid f^{(k)}(x) - P_n^{(k)}(f,x) \mid \leq C_{r,k} \left(\delta_n(x)\right)^{r-k} \omega\left(f^{(r)}, \delta_n(x)\right),$$

where the constant $C_{r,k}$ does not depend on f or n.

There are other similar inequalities due to Gonska and Hinnemann.

Theorem 2.8.13 (Gonska and Hinnemann, [147]). *Fix an integer $r \geq 0$, a constant
C_r and let $L_n : C[-1,1] \to \mathbb{P}_n$ ($n \geq r$) be a sequence of linear operators such that,
for every $x \in [-1,1]$ and $f \in C^r[-1,1]$,*

(i) $\|L_n(f)\| \leq C_r \|f\|, \ f \in C[-1,1],$
(ii) $\mid f(x) - L_n(f,x) \mid \leq C_r \left(\Delta_n(x)\right)^r \|f^{(r)}\|.$

Then, there exists a constant D_r such that, for each $0 \leq k \leq r$ and $f \in C^r[-1,1]$,

$$\|L_n^{(k)}(f)\| \leq C_r \|f^{(k)}\|.$$

Theorem 2.8.14 ([147]). *Fix $r \geq 0$, $s \geq 1$ and let C_r and $C_{r,s}$ be constants.*

(i) *There exists a constant D_r such that, if $f \in C^r[-1,1]$ and $P_n \in \mathbb{P}_n$ ($n \geq r$)
satisfies*

$$\mid f(x) - P_n(x) \mid \leq C_r \left(\Delta_n(x)\right)^r \|f^{(r)}\|,$$

then for $0 \leq k \leq r$

$$\mid f^{(k)}(x) - P_n^{(k)}(x) \mid \leq D_r \left(\Delta_n(x)\right)^{r-k} \|f^{(r)}\|.$$

(ii) *There exists a constant $M_{r,s}$ such that, if $f \in C^r[-1,1]$ and for $P_n \in \mathbb{P}_n$ $(n \geq r + s)$ one has*

$$| f(x) - P_n(x) | \leq C_{r,s} (\Delta_n(x))^r \omega_s(f^{(r)}, \Delta_n(x)),$$

then, for $0 \leq k \leq r$,

$$| f^{(k)}(x) - P_n^{(k)}(x) | \leq M_{r,s} (\Delta_n(x))^{r-k} \omega_s(f^{(r)}, \Delta_n(x)).$$

The Hasson results (Theorem 2.8.5 and 2.8.6) involve estimates in norm. In [234] Leviatan found point-wise estimates in the spirit of the results of Timan and Trigub, but considering the best approximation instead of the modulus of smoothness of the derivatives.

Theorem 2.8.15 (Leviatan, [234]). *For $r \geq 0$ let $f \in C^r[-1,1]$ and let $P_n \in \mathbb{P}_n$ denote its polynomial of best approximation on $[-1,1]$. Then for each $0 \leq k \leq r$ and every $-1 \leq x \leq 1$,*

$$| f^{(k)}(x) - P_n^{(k)}(x) | \leq \frac{C_r}{n^r} [\Delta_n(x)]^{-k} E_{n-k}(f^{(k)}), \quad n \geq k,$$

and

$$| f^{(k)}(x) - P_n^{(k)}(x) | \leq \frac{C_r}{n^r} [\Delta_n(x)]^{-k} E_{n-r}(f^{(r)}), \quad n \geq k,$$

where C_r is an absolute constant which depends only on r.

Proof. For $k = 0$ the result is evident. Assume that it is true for r. By induction, for $f \in C^{r+1}[0,1]$ one has

$$| f^{(k+1)}(x) - Q_{n-1}^{(k)}(x) | \leq \frac{M}{n^k} (\Delta_n(x))^{-k} E_{n-1-k}(f^{(k+1)}), \quad 0 \leq k \leq r, \quad n \geq k,$$

where Q_{n-1} is the polynomial of best approximation to f'. If we set

$$g(x) = f(x) - \int_{-1}^{x} Q_{n-1}(t) dt = f(x) - Q_n(x), \quad x \in [-1,1],$$

then, $| g'(x)) | \leq C E_{n-1}(f')$.

There exists a polynomial S_n such that

$$\|g - S_n\| \leq \frac{C}{n} E_{n-1}(f'), \quad \text{and} \quad \|S_n'\| \leq C E_{n-1}(f').$$

Thus, from (iii) of Theorem 2.7.4 one has

$$| S_n^{(k)}(x) | \leq C(\Delta_n(x))^{1-k} \|S_n'\| \leq C_1(\Delta_n(x))^{1-k} E_{n-1}(f').$$

Let R_n be the polynomial of best approximation to g. Using again (iii) of Theorem 2.7.4 and taking into account that $E_n(g) = E_n(f)$, one has

$$| R_n^{(k)}(x) - S_n^{(k)}(x) | \le C(\Delta_n(x))^{-k} \| R_n - S_n \| \le C(\Delta_n(x))^{-k} \left(E_n(g) + \| g - S_n \| \right)$$

$$\le C_1 \frac{(\Delta_n(x))^{-k}}{n} E_{n-1}(f').$$

Therefore

$$| R_n^{(k)}(x) | \le C_2 \frac{(\Delta_n(x))^{-k}}{n} E_{n-1}(f') \le C_3 \frac{(\Delta_n(x))^{-k}}{n^k} E_{n-k}(f^{(k)}).$$

Since $P_n = Q_n + R_n$ is the polynomial of the best approximation of f we have the result. $\qquad\square$

For the last theorem some of the results of Hasson are easily derived.

Theorem 2.8.16 ([234]). *For $r \ge 0$ let $f \in C^r[-1,1]$ and let $n \ge r$. Then there exists a polynomial $P_n \in \mathbb{P}_n$ such that*

$$| f^{(k)}(x) - P_n^{(k)}(x) | \le C_r \, [\Delta_n(x)]^{r-k} E_{n-r}(f^{(r)}), \quad n \ge k, \tag{2.39}$$

for $k = 0, 1, \ldots, r$ and $-1 \le x \le 1$.

Kilgore combined the estimates of Gopengauz and Leviatan.

Theorem 2.8.17 (Kilgore, [191]). *If $f \in C^m[-1,1]$, for each $n > 2m$, there exists a polynomial $P_n \in \mathbb{P}_n$ such that, for $k = 0, 1, \ldots, m$,*

$$| f^{(k)}(x) - P_n^{(k)}(x) | \le C(m,k) \left(\frac{\sqrt{1-x^2}}{n} \right)^{m-k} E_{n-m}(f^{(m)}), \tag{2.40}$$

where the constants $C(m,k)$ depend only on m and k.

An algebraic analog of the result of Czipszer and Freud in [78] is the following.

Theorem 2.8.18 (Kilgore and Szabados, [196]). *Let $g \in C^q[-1,1]$ be such that $g^{(k)}(\pm 1) = 0$ for $k \le q - 1$. Let $\varepsilon > 0$ and assume there is a sequence $\{P_{n+q}\}$ $(P_{n+q} \in \mathbb{P}_{n+q})$ such that*

$$\left| \frac{g(x) - P_{n+q}(x)}{(\sqrt{1-x^2})^q} \right| \le \frac{\varepsilon}{n^q}.$$

Then, for $|x| \le 1$ and $k \le q$,

$$\left| (g(x) - p_n(x))^{(k)} \right| \le \left(\frac{\sqrt{1-x^2}}{n} + \frac{1}{n^2} \right)^{q-k} \left(\delta_{k,q} \inf_{p_n} \| (g - p_n)^{(q)} \| + \gamma_{k,q} \varepsilon \right),$$

where $\delta_{k,q}$ and $\gamma_{k,q}\varepsilon$ depend on k and q.

2.9   Zamansky-type estimates

As we see in Theorem 1.2.1 concerning trigonometric approximation, for $\sigma < s$, the conditions $E_n^*(f) = \|f - T_n\| = \mathcal{O}(n^{-\sigma})$ and $\|T_n^{(s)}\| = \mathcal{O}(n^{-(\sigma - s)})$ are equivalent. As Hasson showed there is not a direct analogue in the algebraic case.

Theorem 2.9.1 (Hasson, [156]). *There exists a function $f \in C[-1, 1]$ such that $E_n(f) \le K/n$ and, if P_n is the polynomial of best approximation to f on $[-1, 1]$, $\|P_n'\|_{[a,b]} > K \log n$, $n \in \mathbb{N}$, whenever $-1 < a < b < 1$.*

Leviatan also studied the growth of the sequence $\{P_n^{(k)}\}$. His proof is based in a theorem of Runck.

Theorem 2.9.2 (Runck, [315]). *For $r \ge 0$ let $f \in C^r[-1, 1]$ and let $n \ge r$. Then there exists a polynomial $P_n \in \mathbb{P}_n$ such that*

$$| f^{(k)}(x) - P_n^{(k)}(x) | \le C_k [\Delta_n(x)]^{r-k} \omega(f^{(r)}, \Delta_n(x)), \qquad 0 \le k \le r$$

and

$$| P_n^{(k)}(x) | \le C_r [\Delta_n(x)]^{r-k} \omega \left(f^{(r)}, \Delta_n(x) \right), \quad k \ge r + 1,$$

with constant independent of f.

Theorem 2.9.3 (Leviatan, [234]). *For $r \ge 0$ let $f \in C^r[-1, 1]$ and let $P_n \in \mathbb{P}_n$, denote its polynomial of best approximation on $[-1, 1]$. Then for each $k > r$ there exists a constant K, depending only on k, such that, for every $-1 \le x \le 1$,*

$$| P_n^{(k)}(x) | \le \frac{K}{n^r} [\Delta_n(x)]^{-k} \omega \left(f^{(r)}, \frac{1}{n} \right), \quad n \in \mathbb{N}.$$

This improves some results of Hasson. In particular, for $k > r$.

$$\|P_n^{(k)}\| \le K \, n^{2k-r} \omega \left(f^{(r)}, \frac{1}{n} \right),$$

where the constant K depends only on k. An extension to an estimate with higher-order moduli is given as follows.

Theorem 2.9.4 ([234]). *For $r \ge 1$, let $f \in C[-1, 1]$ and let $P_n \in \mathbb{P}_n$, denote its polynomial of best approximation on $[-1, 1]$. Then for each $k \ge r$ there exists a constant K depending on k and r, such that for every $-1 \le x \le 1$,*

$$| P_n^{(k)}(x) | \le K [\Delta_n(x)]^{-k} \omega_r \left(f, \frac{1}{n} \right) \quad n \in \mathbb{N}. \tag{2.41}$$

There is also a nice remark of Leviatan in the paper quoted above: the upper bound of the K-functional in the characterization of the usual modulus of continuity can be given by polynomials. That is, for every $f \in C[-1, 1]$ and $n \in \mathbb{N}$

there is a polynomial $P_n \in \mathbb{P}_n$ such that

$$\|f - P_n\| \leq C\, \omega_r\left(f, \frac{1}{n}\right) \quad \text{and} \quad \|P_n^{(r)}\| \leq C\, n^r\, \omega_r\left(f, \frac{1}{n}\right).$$

In 1985 Ditzian [95] improved (2.41) by proving a similar inequality but in terms of so-called Ditzian-Totik moduli (the definition will be given below). That is

$$|\, P_n^{(k)}(x)\,| \leq K\,[\Delta_n(x)]^{-k}\, \omega_r^\varphi\left(f, \frac{1}{n}\right).$$

The results are the best possible. If $|\, P_n^{(k)}(x)\,| \leq K\,[\Delta_n(x)]^{-k}\, \psi(n)$ where $\psi(n)$ is decreasing, $\psi(n) = o(1)$, and satisfies some additional conditions, then $\omega_r^\varphi(f, 1/n) \leq M\psi(n)$. This provides the analogue to the Sunouchi-Zamanski theorem.

Theorem 2.9.5 (Ditzian, [95]). *If for some integer r and decreasing sequence $\psi(n)$,*

$$\sum_{k=1}^{l} 2^{kr}\psi(2^k) \leq M\, 2^{lr}\psi(2^l) \quad \text{and} \quad E_n(f) \leq \psi(n),$$

then for P_n, the polynomial satisfying $\|f - P_n\| = E_n(f)$, one has

$$|\, P_n^{(k)}(x)\,| \leq K\,[\Delta_n(x)]^{-k}\, \psi(n).$$

In particular, if for some r,

$$\sum_{k=1}^{l} 2^{kr} E_{2^k}(f) \leq M\, 2^{lr} E_{2^l}(f)$$

then

$$|\, P_n^{(k)}(x)\,| \leq K\,[\Delta_n(x)]^{-k}\, E_n(f).$$

Another extension is due to Shevchuk.

Theorem 2.9.6 (Shevchuk, [338]). *If $f \in C^r[-1,1]$ and $\omega_k(f^{(r)}, t) \leq \omega(t)$ ($0 < t \leq 1/k$), then for any $n \geq r + k - 1$ there exists $P_n \in \mathbb{P}_n$ such that, for all $x \in [-1,1]$,*

$$|\, f^{(j)}(x) - P_n^{(j)}(x)\,| \leq C(\Delta_n(x))^{r-j}\, \omega(\Delta_n(x)), \qquad 0 \leq j \leq r,$$

and

$$|\, P_n^{(j)}(x)\,| \leq C(\Delta_n(x))^{r-j}\, \omega(\Delta_n(x)) + C(r + k - j)(\Delta_n(x))^{-j}\, \|f\|_{x,n},$$

for $0 \leq j \leq r + k$, where

$$\|f\|_{x,n} = \max\{|\, f(u)\,| : u \in [x - \Delta_n(x), x + \Delta_n(x)] \cap [-1,1]\}.$$

2.10   Fuksman-Potapov solution to the second problem

In the last sections we have seen theorems which provide characterization for certain classes of functions. For instance, Theorem 2.7.3 characterizes functions satisfying $\omega(f^{(r)}, t) \le C\omega(t)$ by means of its approximations for algebraic polynomials. In Theorems 2.7.5 and 2.7.7 similar results were presented for functions satisfying $f^{(r)} \in \text{Lip}_\alpha[-1, 1]$ or in the Zygmund class respectively. These theorems provide the analogue of the first interpretation after (1.11). That is, we have a characterization of functions satisfying a classical Lipschitz condition in terms of the rate of pointwise approximation by algebraic polynomials. Let us consider the problem of characterization of other classes of functions.

For $r \in \mathbb{N}$ and $\alpha \in (0, 1)$, let

$$K(r, \alpha) = \{f \in C[-1, 1]: \; E_n(f) \le M(f) n^{-r-\alpha}\}.$$

Classes $K(r, \alpha)$ are defined in terms of the rate of convergence of the best approximation. The classes $C^{r,\alpha}[-1, 1]$ and $K(r, \alpha)$ are different. For instance, for $f(x) = \sqrt{1 - x^2}$ one has, $f \in K(0, 1)$ but, for any $\delta > 1/2$, $f \notin C^{0,\delta}[-1, 1]$.

It was an interesting question to describe classes $K(r, \alpha)$ without any reference to approximation by polynomials. One of the first results in this direction is due to Fuksman [129]. For $f \in C^r(-1, 1)$ and $0 \le k \le r/2$, let $\psi_k(x) = f^{(r-k)}(x)(1 - x^2)^{r/2-k}$ and consider the condition

$$\sup_{h \,\in\, \Lambda(x, \delta)} | \psi_k(x) - \psi_k(x + h) | \le C \left(\frac{\delta}{\sqrt{1 - x^2} + \sqrt{\delta}} \right)^\alpha, \qquad (2.42)$$

where $\Lambda(x, \delta) = \{h :| h \,|\le \delta, | \, x + h \,|\le 1\}$. We assume $\psi_1(1) = \psi_1(-1) = 0$ for odd k. Let

$$S(r, \alpha) = \{f \in C^r(-1, 1) : \psi_k \in C[-1, 1] \; (0 \le k \le r/2) \quad \text{and} \quad (2.42) \quad \text{holds}\,\}.$$

Theorem 2.10.1 (Fuksman, [129]). *For each $r \in \mathbb{N}_0$ and $0 < \alpha < 1$, one has $K(r, \alpha) = S(r, \alpha)$.*

Proof. In order to verify the inclusion $S(r, \alpha) \subset K(r, \alpha)$, for $f \in S(r, \alpha)$, define $F(t) = f(\cos(t))$.

If $r = 0$, then $\psi_0(x) = f(x)$ and (2.42) yields

$$| f(x + h) - f(x) | \le C \left(\frac{\delta}{\sqrt{1 - x^2} + \sqrt{\delta}} \right)^\alpha \le C \min \left\{ \left(\frac{\delta}{\sqrt{1 - x^2}} \right)^\alpha, \left(\sqrt{\delta} \right)^\alpha \right\}$$

for $h \in \lambda(x, \delta)$. Set $h = \cos(t + h) - \cos t$ and $\delta =| h \sin t | + h^2$. Since

$$| \cos(t + h) - \cos t |=| \cos t(1 - \cos h) + \sin t \sin h |\le \delta,$$

one has

$$| F(t+h) - F(t) | \leq C \min \left\{ \left(\frac{\delta}{\sqrt{1 - \cos^2 t}} \right)^{\alpha}, \left(\sqrt{\delta} \right)^{\alpha} \right\}. \qquad (2.43)$$

We should consider two cases.

Case 1. If $| h | \leq | \sin t |$, then

$$\frac{\delta}{| \sin t |} = \frac{| h \sin t | + h^2}{| \sin t |} \leq 2 | h |.$$

Case 2. If $| h | > | \sin t |$, then

$$\sqrt{\delta} = \sqrt{| h \sin t | + h^2} \leq | h | \sqrt{2} \leq 2 | h |.$$

Therefore

$$| F(t+h) - F(t) | \leq C_1 | h |^{\alpha}.$$

Now we consider that $r > 0$. By induction with respect to r it can be proved that there exists trigonometric polynomials $\varphi_{i,r} \in \mathbb{T}_{r-i}$ such that

$$F^{(r)}(t) = \sum_{i=0}^{[r/2]} f^{(r-i)}(\cos t) \sin^{r-2i}(t) \varphi_{i,k}(t) + \sum_{i=[r/2]+1}^{r} f^{(r-i)}(\cos t) \varphi_{i,k}(t).$$

But

$$f^{(r-i)}(\cos t) | \sin^{r-2i}(t) | = \psi_i(\cos t),$$

then we can write

$$F^{(r)}(t) = \sum_{i=0}^{[r/2]} \Psi_i(t) \varphi_{i,k}(t) + \sum_{i=[r/2]+1}^{r} f^{(r-i)}(\cos t) \varphi_{i,k}(t),$$

where $\Psi_i(t) = f^{(r-i)}(\cos t)\text{sign}(\sin t)^k$. It can be proved that these functions are continuous. Moreover, as in the proof of the case $r = 0$, each function Ψ_i satisfies a Lipschitz condition of order α. Therefore, there exist a constant C and a sequence $\{T_n\}$ of even trigonometric polynomials such that $| F(t) - T_n(t) | \leq Cn^{-(k+\alpha)}$. By taking $P_n(x) = T_n(\arccos x)$ we conclude that $f \in K(r, \alpha)$.

Let us consider the relation $K(r, \alpha) \subset S(r, \alpha)$. Fix $f \in K(r, \alpha)$ and a sequence $\{P_n\}$ of polynomials such that $\|f - P_n\| \leq Cn^{-(k+\alpha)}$. If we set

$$Q_n = P_{2^n} - P_{2^{n-1}} \qquad (n \geq 1), \qquad (2.44)$$

then

$$\|Q_n\| \leq C2^{-n(r+\alpha)} \qquad (2.45)$$

and take into account that

$$f(x) = \sum_{n=1}^{\infty} Q_n(x),$$

then

$$f^{(i)}(x) = \sum_{n=1}^{\infty} Q_n^{(i)}(x), \qquad (i = 0, 1, \ldots, r).$$

Set $\psi_j(x) = f^{(r-j)}(x)(1-x^2)^{r/2-j}$, then

$$\psi_j(x) = \sum_{n=0}^{\infty} Q_n^{(r-j)}(x)(1-x^2)^{r/2-j} = \sum_{n=0}^{m} + \sum_{n=m+1}^{\infty} = L_m(x) + L_m^*(x), \quad (2.46)$$

where m will chosen later.

We will estimate the modulus of continuity of L_m and L_m^*. First

$$|L_m(x+h) - L_m(x)| \le |h| \sum_{n=0}^{m} \left| \frac{d}{du} \left[(1-u^2)^{r/2-j} Q_n^{r-j}(u) \right] \right|_{u=x+h\theta}$$

$$\le |h| \sum_{n=0}^{m} \left\{ \left| 2u(1-u^2)^{r/2-j-1} Q_n^{r-j}(u) \right| + \left| (1-u^2)^{r/2-j} Q_n^{r-j-1}(u) \right| \right\}_{u=x+h\theta}.$$

Now we have two different estimates: taking into account (2.45) and (2.23) (with $l = r - j$, $q = r - 2j - 1$ $p = j + 1$ $(p + q = l)$) one has

$$\sum_{n=0}^{m} 2u(1-u^2)^{r/2-j-1} Q_n^{r-j}(u) \Big|_{u=x+h\theta}$$

$$\le C_1 \sum_{n=0}^{m} 2^{(r+1)n} \|Q_n\| (1-u^2)^{r/2-j-1} (1-u^2)^{-r/2+j+1/2} \Big|_{u=x+h\theta}$$

$$\le C_2 (1 - (x+h\theta)^2))^{-1/2} \sum_{n=0}^{m} 2^{(r+1)n} 2^{-n(r+\alpha)} \le C_3 \frac{2^{(1-\alpha)m}}{(1 - (x+h\theta)^2))^{1/2}}.$$

On the other hand, (2.23) (with $l = r - j$, $q = r - 2j - 2$ $p = j + 2$ $(p + q = l)$) one has

$$\sum_{n=0}^{m} 2u(1-u^2)^{r/2-j-1} Q_n^{r-j}(u) \Big|_{u=x+h\theta}$$

$$\le C_1 \sum_{n=0}^{m} 2^{(r+2)n} \|Q_n\| (1-u^2)^{r/2-j-1} (1-u^2)^{-r/2+j+1} \Big|_{u=x+h\theta}$$

$$\le C_2 \sum_{n=0}^{m} 2^{(r+2)n} 2^{-n(r+\alpha)} \le C_3 2^{(2-\alpha)m}.$$

Since for the other term in the estimate of $| L_m(x+h) - L_m(x) |$ we can obtain similar inequalities, we have proved that

$$| L_m(x+h) - L_m(x) | \le C_4 \frac{2^{(1-\alpha)m}}{(1-(x+h\theta)^2))^{1/2}} \tag{2.47}$$

and

$$| L_m(x+h) - L_m(x) | \le C_5 \, 2^{(2-\alpha)m}. \tag{2.48}$$

With similar arguments we also prove that

$$| L_m^*(x+h) - L_m^*(x) | \le C_6 \, 2^{-\alpha m}. \tag{2.49}$$

If $\varepsilon > 0$, $| h | \le \varepsilon$ and $| x | \le 1-\varepsilon$, we take m such that $2^m < \sqrt{(1-\varepsilon)^2 - x^2} = r \le 2^{m+1}$. Then from (2.46), (2.47) and (2.49) we obtain

$$| \psi_j(x+h) - \psi_j(x) | \le C \left(| h | \left(\frac{| h |}{r} \right)^{1-\alpha} \frac{1}{r} + \left(\frac{r}{| h |} \right)^{\alpha} \right) = 2C \left(\frac{r}{| h |} \right)^{\alpha}.$$

If we take m such that $2^m < (| h |)^{-1/2} \le 2^{m+1}$, then from (2.46), (2.48) and (2.49) we obtain

$$| \psi_j(x+h) - \psi_j(x) | \le C \left(| h | \left(\frac{1}{\sqrt{| h |}} \right)^{2-\alpha} + \left(\frac{1}{\sqrt{| h |}} \right)^{\alpha} \right) = 2C \, (| h |)^{\alpha/2}.$$

Thus, if $h \in \Lambda(x, \varepsilon)$, then

$$| \psi_j(x+h) - \psi_j(x) | \le C \min \left((| h | / r)^{\alpha}, | h |^{\alpha/2} \right)$$
$$= C | h |^{\alpha} \min \left((1/r), | h |^{-1/2} \right)^{\alpha}$$
$$\le \frac{C_1 | h |^{\alpha}}{(r + \sqrt{| h |})^{\alpha}} = C_1 \left(\frac{| h |}{r + \sqrt{| h |}} \right)^{\alpha}$$
$$\le C_1 \left(\frac{\varepsilon}{r + \sqrt{\varepsilon}} \right)^{\alpha} \le C_2 \left(\frac{\varepsilon}{\sqrt{1 - x^2} + \sqrt{\varepsilon}} \right)^{\alpha},$$

since

$$\frac{1}{r + \sqrt{\varepsilon}} = \frac{1}{\sqrt{(1-\varepsilon)^2 - x^2} + \sqrt{\varepsilon}} \le \frac{6}{\sqrt{1 - x^2} + \sqrt{\varepsilon}}.$$

Finally, if r is odd, from (2.49) we know that the series (2.46) converges uniformly on $[-1, 1]$. Moreover, for $j < [r/2]$, one has $k > 2j$, thus $\psi_j(\pm 1) = 0$. We have proved that $f \in S(r, \alpha)$. $\quad\square$

The result also can be extended to the case when

$$E_n(f) \le \frac{M(f)}{n^k} \, \omega \left(\frac{1}{n} \right)$$

where ω is a modulus of continuity satisfying conditions (2.22). In this case, the definition of the class $S(r,\omega)$ is similar to the one of $S(r,k)$, but condition (2.42) is replaced by

$$\sup_{(h,x)\,\in\,\Lambda(\delta)} |\,\psi_k(x) - \psi_k(x+h)\,| \le C\omega\left(\frac{\delta}{\sqrt{1-x^2}+\sqrt{\delta}}\right).$$

Theorem 2.10.2 (Fuksman, [129]). *Let $f \in C[-1,1]$, r a positive integer and $\alpha \in (0,1)$. The following assertions are equivalent:*

(i) *$f \in C^{2r}(-1,1)$ and, for $0 \le k \le r$ and $\psi_k(x) = f^{(r-k)}(x)(1-x^2)^{r-k}$, one has $\psi_k \in C[-1,1]$ and (2.42) holds.*

(ii) *For each $n \in \mathbb{N}$ there exists an algebraic polynomial $P_n \in \Pi_{n-1}$ such that, for each $x \in [-1,1]$,*

$$|\,f(x) - P_n(x)\,| \le \frac{D}{n^{2r+\alpha}}$$

where D is a positive constant which does not depend on x or n.

In 1980 Potapov [295] unified the results of Dzyadyk and Fuksman. He proved an analogue to Theorem 2.10.3, but with the condition $\alpha + \beta < 1$ instead of $\alpha + \beta/2 < 1$. Notice that, by taking $\beta = 0$ we obtain the Dzyadyk characterization and for $\beta = -\alpha$ the Fuksman result. The results we present here were proved by Potapov in 2005 [303].

Theorem 2.10.3 (Potapov, [303]). *Fix reals α and β such that $\alpha \in (0,1)$, $\alpha+\beta \ge 0$ and $\alpha + \beta/2 < 1$. For $f \in C[-1,1]$ the following assertions are equivalent:*

(i) *For each $x \in [-1,1]$, one has*

$$\sup_{\{h\,:\,|h|\le\delta,\,|x+h|\le1\}} |\,f(x+h) - f(x)\,| \le C_1\,\delta^\alpha\left(\sqrt{1-x^2}+\sqrt{\delta}\right)^\beta,$$

where C_1 is a positive constant which does not depend on δ or x;

(ii) *For each $n \in \mathbb{N}$ there exists $P_{n-1} \in \Pi_{n-1}$ such that, for each $x \in [-1,1]$,*

$$|\,f(x) - P_{n-1}(x)\,| \le C_2\,n^\beta\,(\Delta_n(x))^{\alpha+\beta}$$

where C_2 is a positive constant which does not depend on x or n.

Proof. (i) $\Longrightarrow$ (ii). Fix $m, s \in \mathbb{N}$ such that $(n-1)/s < m \le 1 + (n-1)/s$ and define

$$Q(x) = \int_{-\pi}^{\pi} f(\cos(t+y))K_{m,s}(t)dt,$$

where $x = \cos y$ and where K_{2q} is given by (2.8) with $s = q$. There exist positive constant C_1 and C_2 such that

$$C_1m^{2s-1} \le c_{m,s} \le C_2m^{2s-1} \quad \text{and} \quad C_1m^\beta \le \int_{-\pi}^{\pi} |\,t\,|^\beta\,K_{m,s}(t)dt \le C_2m^{-\beta}.$$

Moreover, $Q_m \in \mathbb{P}_{n-1}$.

Since for $|t| \leq \pi$, one has

$$\gamma = |t| \sqrt{1-x^2} + t^2)^\alpha \leq 3(|t| \sqrt{1-x^2} + t^2),$$

as in the proof of (2.43) from (i) we have

$$|f(\cos(y+t)) - f(\cos y)| \leq C (|t| \sqrt{1-x^2} + t^2)^\alpha (\gamma)^\beta$$
$$\leq 3C |t|^\alpha (\sqrt{1-x^2} + t^2)^{\alpha+\beta}$$
$$\leq C_1 (|t|^\alpha (\sqrt{1-x^2})^{\alpha+\beta} + |t|^{2\alpha+\beta}).$$

Now one has

$$|f(x) - Q(x)| \leq \int_{-\pi}^{\pi} |f(\cos(t+y)) - f(\cos y)| K_{m,s}(t)dt$$
$$\leq C \left((\sqrt{1-x^2})^{\alpha+\beta} \int_{-\pi}^{\pi} |t|^\alpha K_{m,s}(t)dt + \int_{-\pi}^{\pi} |t|^{2\alpha+\beta} K_{m,s}(t)dt \right)$$
$$\leq C_1 \left(\frac{(\sqrt{1-x^2})^{\alpha+\beta}}{m^\alpha} + \frac{1}{m^{2\alpha+\beta}} \right) \leq C_2 n^\beta (\Delta_n(x))^{\alpha+\beta}.$$

We have proved (ii).

(ii) $\Longrightarrow$ (i). We should modify the arguments of the proof of Theorem 2.10.1. If Q_n be defined by (2.44), then

$$|Q_k(x)| \leq C2^{k\beta} (\Delta_{2^k}(x))^{\alpha+\beta}$$

and from Theorem 2.7.5 we obtain

$$|Q_k(x+h) - Q_k(x)| \leq C_1 |h| 2^{k\beta} (\Delta_{2^k}(x+h\theta))^{\alpha+\beta-1}.$$

Fix $x \in [-1,1]$ and $|x+h| \leq 1$. Fix $N \in \mathbb{N}$ which will be chosen later. Notice that

$$|\Delta_h f(x)|$$

$$\leq |f(x) - P_{2^N}(x)| + |f(x+h) - P_{2^N}(x+h)| + \sum_{k=0}^{N} |Q_k(x+h) - Q_k(x)|$$

$$\leq C_3 \left(2^{N\beta}((\Delta_{2^N}(x))^{\alpha+\beta} + (\Delta_{2^N}(x+h))^{\alpha+\beta}) + |h| \sum_{k=0}^{N} \frac{(\Delta_{2^k}(x+h\theta))^{\alpha+\beta-1}}{2^{-k\beta}} \right)$$

$$\leq C_3 2^{N\beta} ((\Delta_{2^N}(x))^{\alpha+\beta} + (\Delta_{2^N}(x+h))^{\alpha+\beta}) + |h| (\Delta_{2^N}(x+h\theta))^{\alpha+\beta-1}),$$

where the sum is estimated as follows. If $\alpha + \beta \leq 1$ and $\alpha < 1$, then

$$2^{k(\alpha+\beta-1)}(\Delta_{2^k}(x+h\theta))^{\alpha+\beta-1} \leq 2^{N(\alpha+\beta-1)}(\Delta_{2^N}(x+h\theta))^{\alpha+\beta-1}.$$

Hence

$$\sum_{k=0}^{N} \frac{(\Delta_{2^k}(x+h\theta))^{\alpha+\beta-1}}{2^{-k\beta}} \le 2^{N(\alpha+\beta-1)}(\Delta_{2^N}(x+h\theta))^{\alpha+\beta-1}\sum_{k=0}^{N} 2^{k(1-\alpha)}$$

$$\le C\, 2^{N\beta}(\Delta_{2^N}(x+h\theta))^{\alpha+\beta-1}.$$

On the other hand, if $\alpha+\beta > 1$ and $\alpha+\beta/2 < 1$, then

$$2^{k(\alpha+\beta-1)}(\Delta_{2^k}(x+h\theta))^{\alpha+\beta-1} \le \left(\frac{2^N}{2^k}\right)^{\alpha+\beta-1} 2^{N(\alpha+\beta-1)}(\Delta_{2^N}(x+h\theta))^{\alpha+\beta-1}.$$

Hence

$$\sum_{k=0}^{N} \frac{(\Delta_{2^k}(x+h\theta))^{\alpha+\beta-1}}{2^{-k\beta}} \le 2^{2N(\alpha+\beta-1)}(\Delta_{2^N}(x+h\theta))^{\alpha+\beta-1}\sum_{k=0}^{N} 2^{k(2-2\alpha-\beta)}$$

$$\le C\, 2^{2N(\alpha+\beta-1)}(\Delta_{2^N}(x+h\theta))^{\alpha+\beta-1}2^{N(2-2\alpha-\beta)}$$

$$= C\, 2^{N\beta}(\Delta_{2^N}(x+h\theta))^{\alpha+\beta-1}.$$

To finish the proof we should choose N.

Case 1. Suppose that $0 < h < 1/4$ and $x \in [-1, -1+2h] \cup [1-2h, 1-h]$. Chose N such that $2^{-2N-1} \le h < 2^{-2N}$. Then

$$1 - x^2 \le 1 - (1-2h)^2 \le 4h \le 4\, 2^{-2N}$$

and

$$1 - (x+h\theta)^2 \le 1 - x^2 + 2h \le 6h \le 6\, 2^{-2N}.$$

Hence

$$2^{-N} \le \sqrt{1-x^2} + 2^{-N} \le 3 2^{-N},$$

$$2^{-N} \le \sqrt{1-(x+h)^2} + 2^{-N} \le 4 2^{-N},$$

$$2^{-N} \le \sqrt{1-(x+h\theta)^2} + 2^{-N} \le 4\, 2^{-N},$$

and

$$\frac{1}{3}\left(\sqrt{h} + \sqrt{1-x^2}\right) \le \sqrt{h} \le \sqrt{h} + \sqrt{1-x^2},$$

and we obtain

$$|\, f(x+h) - f(x)\,| \le \frac{C}{2^{N\alpha}}(2^{-N(\alpha+\beta)} + h2^N 2^{-N(\alpha+\beta-1)})$$

$$\le C_1\, 2^{-N(2\alpha+\beta)} \le C_2\, h^{\alpha+\beta/2} \le C_3\, h^{\alpha}(\sqrt{1-x^2} + \sqrt{h})^{\beta}.$$

Case 2. Suppose that $0 < h < 1/4$ and $x \in [-1+2h, 1-2h]$. Choose N such that

$$\frac{\sqrt{1-x^2}}{2^{N+1}} < h \le \frac{\sqrt{1-x^2}}{2^N}.$$

Now we consider the inequalities

$$2\sqrt{h} \le \sqrt{1 - (1 - 2h)^2} \le \sqrt{1 - x^2},$$

$$2\sqrt{1 - x^2} \le \sqrt{1 - x^2} + \frac{2h}{\sqrt{1 - x^2}} \le \sqrt{1 - x^2} + \frac{1}{2^N} < \sqrt{1 - x^2},$$

$$1 - (x + h\theta)^2 \le 1 - x^2 + 4h \le 2(1 - x^2),$$

and

$$1 - x^2 = 1 - (x + h\theta)^2 + h\theta(2x + h\theta) \le 1 - (x + h\theta)^2 + 2h \le 1 - (x + h\theta)^2 + 2\frac{\sqrt{1 - x^2}}{2^N}.$$

Then

$$\left(\sqrt{1 - x^2} - \frac{1}{2^N}\right)^2 = 1 - x^2 - 2\frac{\sqrt{1 - x^2}}{2^N}$$

$$\le 1 - (x + h\theta)^2 + \frac{1}{2^N} \le \left(\sqrt{1 - (x + h\theta)^2} + \frac{1}{2^N}\right)^2.$$

Therefore, if $2^{-N} \le \sqrt{1 - x^2}$, then

$$\sqrt{1 - x^2} - \frac{1}{2^N} \le \sqrt{1 - (x + h\theta)^2} + \frac{1}{2^N},$$

and

$$\sqrt{1 - x^2} \le 2(\sqrt{1 - (x + h\theta)^2} + \frac{1}{2^N}.$$

On the other hand, if $2^{-N} > \sqrt{1 - x^2}$, then

$$\sqrt{1 - x^2} \le \sqrt{1 - (x + h\theta)^2} + \frac{1}{2^N}.$$

Hence, in this case

$$\frac{1}{2}\sqrt{1 - x^2} \le \sqrt{1 - (x + h\theta)^2} + \frac{1}{2^N} \le 2\left(\sqrt{1 - x^2} + \frac{1}{2^N}\right) \le 4\sqrt{1 - x^2}.$$

With these inequalities we obtain

$$|f(x + h) - f(x)| \le \frac{C}{2^{N\alpha}}\left(\sqrt{1 - x^2}\,^{\alpha + \beta} + h2^N(\sqrt{1 - x^2})^{\alpha + \beta - 1}\right)$$

$$\le \frac{C_2}{2^{N\alpha}}(\sqrt{1 - x^2})^{\alpha + \beta} \le C_3 h^\alpha(\sqrt{1 - x^2} + \sqrt{h})^\beta.$$

Case 3. The case $h \in (-1/4, 0)$ can be treated as the case $h \in (0, 1/4)$.

Case 4. For $\delta < 1/4$ the proof follows from the arguments given above. For $\delta \ge 1/4$ the proof is simple.   □

The Chebyshev differential operator is defined by

$$D(f, x) = (1 - x^2)f''(x) - xf'(x).$$

Moreover, set $D^{(1)} = D$ and $D^{(r)} = D(D^{(r-1)})$, for $r \geq 2$.

Theorem 2.10.4 (Potapov, [303]). *Fix real numbers $\sigma \geq 0$ and $\gamma > 0$. For $f \in C[-1, 1]$, the following assertions are equivalent:*

(i) *For each $n \in \mathbb{N}$ $(n \geq 2)$ there exists $P_{n-1} \in \Pi_{n-1}$ such that, for each $x \in [-1, 1]$,*

$$|\, f(x) - P_{n-1}(x)\,| \leq \frac{C_1}{n^{2+\gamma-\sigma}} \left(\Delta_n(x)\right)^\sigma$$

where C_1 is a positive constant which does not depend on x or n.

(ii) *For any interval $[a, b] \subset (-1, 1)$, $f \in C^2[a, b]$, $Df \in C[-1, 1]$ and for each $n \in \mathbb{N}$ $(n \geq 2)$ there exists $R_{n-1} \in \Pi_{n-1}$ such that, for each $x \in [-1, 1]$,*

$$|\, Df(x) - R_{n-1}(x)\,| \leq \frac{C_2}{n^{\gamma-\sigma}} \left(\Delta_n(x)\right)^\sigma$$

where C_3 is a positive constant which does not depend on x or n.

(iii) *For any interval $[a, b] \subset (-1, 1)$, $f \in C^2[a, b]$, $f'(x), (1 - x^2)f''(x) \in C[-1, 1]$ and for each $n \in \mathbb{N}$ $(n \geq 2)$ there exists $Q_{n-1,1}, Q_{n-1,2} \in \Pi_{n-1}$ such that, for each $x \in [-1, 1]$,*

$$|\, f'(x) - Q_{n-1,1}(x)\,| \leq C_3\, n^{\sigma-\gamma} \left(\Delta_n(x)\right)^\sigma$$

and

$$|\, (1 - x^2)f''(x) - Q_{n-1,2}(x)\,| \leq C_1\, n^{\sigma-\gamma} \left(\Delta_n(x)\right)^\sigma$$

where C_3 is a positive constant which does not depend on x or n.

Theorem 2.10.5 ([303]). *Fix real numbers $\sigma \geq 0$ and $\gamma > 0$. For $f \in C[-1, 1]$, the following assertions are equivalent:*

(i) *For each $n \in \mathbb{N}$ $(n \geq 2)$ there exists $P_{n-1} \in \Pi_{n-1}$ such that, for each $x \in [-1, 1]$,*

$$|\, f(x) - P_{n-1}(x)\,| \leq \frac{C_1}{n^{\gamma-\sigma}} \left(\Delta_n(x)\right)^{\sigma+1}$$

where C_3 is a positive constant which does not depend on x or n.

(ii) *$f \in C^1[-1, 1]$ and for each $n \in \mathbb{N}$ $(n \geq 2)$ there exists $R_{n-1} \in \Pi_{n-1}$ such that, for each $x \in [-1, 1]$,*

$$|\, f'(x) - R_{n-1}(x)\,| \leq \frac{C_1}{n^{\gamma-\sigma}} \left(\Delta_n(x)\right)^\sigma$$

where C_3 is a positive constant which does not depend on x or n.

Theorem 2.10.6 ([303]). *Let $f \in C[-1,1]$, r and ρ be non-negative integers and fix α and β such that $\alpha \in (0,1)$, $\alpha + \beta \geq 0$ and $\alpha + \beta/2 < 1$. The following assertions are equivalent:*

(i) $f \in C^{2\rho+r}(-1,1)$, $\psi(x) = D^{(\rho)} f^{(r)}(x) \in C[-1,1]$ *and*

$$\sup_{\{\,h\,:\,|h|\leq\delta,\,|x+h|\leq 1\}} |\,\psi(x) - \psi(x+h)\,| \leq C_1 \delta^\alpha \left(\sqrt{1-x^2} + \sqrt{\delta}\right)^\beta,$$

where C_1 is a positive constant which does not depend on δ or x;

(ii) $f \in C^{2\rho+r}(-1,1)$ *and, for $0 \leq k \leq \rho$ and $\psi_k(x) = f^{(2\rho+r-k)}(x)(1-x^2)^{\rho-k}$, one has $\psi_k \in C[-1,1]$ and*

$$\sup_{\{\,h\,:\,|h|\leq\delta,\,|x+h|\leq 1\}} |\,\psi_k(x) - \psi_k(x+h)\,| \leq C_2 \delta^\alpha \left(\sqrt{1-x^2} + \sqrt{\delta}\right)^\beta,$$

where C_2 is a positive constant which does not depend on δ or x;

(iii) *For each $n \in \mathbb{N}$ there exists an algebraic polynomial $P_{n-1} \in \Pi_{n-1}$ such that, for each $x \in [-1,1]$,*

$$|\,f(x) - P_{n-1}(x)\,| \leq \frac{C_3}{n^{2\rho-\beta}} (\Delta_n(x))^{r+\alpha+\beta}$$

where C_3 is a positive constant which does not depend on x or n.

Proof. Assume condition (iii) holds. Then there exists a sequence $\{P_n\}$ of algebraic polynomials for which

$$|\,f(x) - P_{n-1}(x)\,| \leq \frac{C}{n^{2\rho-\beta}} (\Delta_n(x))^{r+\alpha+\beta}.$$

By applying ρ-times Theorem 2.10.5 we obtain that condition (iii) is equivalent to the following condition A: there exists a sequence $\{R_n\}$ ($n \geq 2$) of algebraic polynomials $R_n \in \Pi_n$ such that, for each $x \in [-1,1]$,

$$|\,f^{(r)}(x) - R_n(x)\,| \leq \frac{C}{n^{2\rho-\beta}} (\Delta_n(x))^{\alpha+\beta}.$$

By applying ρ-times Theorem 2.10.4 (which is equivalent to condition (i) and (ii)) we obtain that condition A is equivalent to the following condition B: there exists a sequence $\{Q_n\}$ ($n \geq 2$) of algebraic polynomials $Q_n \in \Pi_n$ such that, for each $x \in [-1,1]$,

$$|\,D^{(\rho)}(f^{(r)}(x)) - Q_n(x)\,| \leq Cn^\beta (\Delta_n(x))^{\alpha+\beta}.$$

Applying Theorem 2.10.3 we obtain that condition B is equivalent to condition (i). Thus we have proved that (i) and (ii) are equivalent.

Let us prove that (ii) and (iii) are equivalent. By applying r-times we obtain that condition (iii) is equivalent to condition A. From Theorem 2.10.4 (condition (i) and (ii) are equivalent) we obtain that condition A is equivalent to condition B: for each $n \in \mathbb{N}$ $(n \geq 2)$ there exist algebraic polynomials $T_{n,1}, T_{n,2} \in \Pi_n$ such that, for each $x \in [-1, 1]$,

$$| f^{(r+1)}(x)(1 - x^2)^{i-1} - T_{n,i}(x) | \leq \frac{C \left(\Delta_n(x)\right)^{\alpha+\beta}}{n^{2(\rho-1)-\beta}}, \qquad i = 1, 2.$$

If $\rho > 1$, then from Theorem 2.10.4 we obtain that condition B is equivalent to the following condition C: for each $n \in \mathbb{N}$ $(n \geq 2)$ there exist algebraic polynomials $H_{n,i} \in \Pi_n$ $(i \in \{1, 2, 3, 4\})$, such that, for each $x \in [-1, 1]$,

$$| f^{(r+1+i)}(x)(1 - x^2)^{i-1} - T_{n,i}(x) | \leq \frac{C \left(\Delta_n(x)\right)^{\alpha+\beta}}{n^{2(\rho-2)-\beta}}, \qquad i = 1, 2$$

and

$$| f^{(r+1+i)}(x)(1 - x^2)^{i-3}(1 - x^2)^{(i-2)} - T_{n,i}(x) | \leq \frac{C \left(\Delta_n(x)\right)^{\alpha+\beta}}{n^{2(\rho-2)-\beta}}, i = 3, 4.$$

It can be proved that condition C is equivalent to the condition D: for each $n \in \mathbb{N}$ $(n \geq 2)$ there exist algebraic polynomials $L_{n,i} \in \Pi_n$ $(i \in \{1, 2, 3\})$, such that, for each $x \in [-1, 1]$,

$$| f^{(r+1+i)}(x)(1 - x^2)^{i-1} - L_{n,i}(x) | \leq \frac{C \left(\Delta_n(x)\right)^{\alpha+\beta}}{n^{2(\rho-2)-\beta}}, \qquad i = 1, 2, 3.$$

If $\rho > 2$, we repeat $\rho - 2$-times the arguments given above to obtain that condition D is equivalent to the following condition E: for each $n \in \mathbb{N}$ $(n \geq 2)$ there exist algebraic polynomials $S_{n,i} \in \Pi_n$ $(i \in \{0, 1, 2, \ldots, \rho\})$, such that, for each $x \in [-1, 1]$,

$$| f^{(r+2\rho-i)}(x)(1 - x^2)^{\rho-i} - S_{n,i}(x) | \leq C n^{\beta} \left(\Delta_n(x)\right)^{\alpha+\beta}, \qquad i = 0, 2, \ldots, \rho.$$

From Theorem 2.10.3 we obtain that condition E is equivalent to condition (ii). Thus we have proved that conditions (ii) and (iii) are equivalent. $\qquad \square$

2.11   Integral metrics

In the works of Timan and Dzyadyk the best approximation by algebraic polynomials was well studied in the case of the uniform norm. Several authors considered that extension of the Timan-type estimates the spaces of integrable functions. The problem of characterization for some classes of functions was considered by Potapov and Lebed.

In 1956 Potapov [287] extended the Timan theorem by considering functions with derivative of order r $(r > 0$, integer) is in $\mathrm{Lip}(p, \alpha)$, $p > 1$, $0 < \alpha \le 1$. He also studied functions satisfying the condition

$$\left(\int_c^d | f^{(r)}(x + h) - f^{(r)}(x) |^p \, \frac{dx}{\sqrt{(x - a)(b - x)}} \right)^{1/p} \le M(f) \mid h \mid^{\alpha},$$

where $a \le c < d \le b$.

In 1958 Lebed obtained a direct result. He considered the term $\Delta_n(x)$ as a varying weight.

Theorem 2.11.1 (Lebed, [226]). *Assume that $p \ge 1$ and $1 - s - 1/p \ge 0$. If $f \in C^m[-1, 1]$ and $\|(\sqrt{1 - x^2})^s f^{(m)}(x)\|_p \le M$, then there exists a sequence $\{P_n\}$ $(P_n \in \mathbb{P}_n)$ such that*

$$\left\| \frac{f(x) - P_n(x)}{(\Delta_n(x))^{m-s}} \right\|_p \le C(m) \frac{M}{n^s}.$$

Denote by $W^{(r)} H_p^w$ the class of functions given on the interval $[-1, 1]$ and having an rth derivative $f^{(r)}$ whose pth power is integrable, and for which the inequality

$$\|f^{(r)}(x + h) - f^{(r)}(x)\|_{L_p[-1, 1-h]} \le w(h), \qquad 0 < h < 1,$$

hods, where w is a fixed modulus of continuity. The class $W^{(r)} A_p^w$ is defined analogously, but with the condition

$$\left\| \frac{f^{(r)}(x + \sqrt{1 - h^2} - h\sqrt{1 - x^2}) - f^{(r)}(x)}{w(h\sqrt{1 - x^2} + h^2)} \right\|_p \le C.$$

For $w(t) = t^{\alpha}$ we shall denote these classes by $H_p^{(r+\alpha)}$ $(A_p^{(r+\alpha)}$ respectively).

The classes $W^{(r)} H_p^w$ were introduced by Lebed and Potapov (see [290]). They proved that $W^{(r)} H_\infty^w = W^{(r)} A_\infty^w$ (uniform norm). It is also obvious that the intersection of these classes is not empty, for $1 \le p < \infty$.

Potapov also used classes defined by two parameters. For $1 \le p < \infty$, $r \in \mathbb{N}_0$, $0 \le \beta \le 1$ and $0 < \alpha \le 1$, $f \in H_p^{(r)} A_\beta^\alpha$ if $f^{(r)} \in L_p[-1, 1]$ if

$$\left(\int_{-1}^1 \left| \frac{f^{(r)}(x\sqrt{1 - h^2} - h\sqrt{1 - x^2}) - f^{(r)}(x)}{\sqrt{1 - x^2} + \mid h \mid^\beta} \right|^p dx \right)^{1/p} \le \mid h \mid^\alpha$$

in the case $0 < \alpha < 1$ and

$$\int_{-1}^1 \left| \frac{f^{(r)}(\lambda(x, h)x - \lambda(h, x)) - 2f^{(r)}(x) + f^{(r)}(\lambda(x, h) + \lambda(h, x))}{\sqrt{1 - x^2} + \mid h \mid^\beta} \right|^p dx \le \mid h \mid^p$$

in the case $\alpha = 1$, where $\lambda(x, h) = x\sqrt{1 - h^2}$. Here is a typical result.

Theorem 2.11.2 (Potapov, [289]). *For a function f one has $f \in H_p^{(r)} A_\beta^\alpha$ if and only if, for each $n \geq r + 2$, there exists a polynomial $P_n \in \mathbb{P}_n$, such that*

$$\left(\int_{-1}^{1} \left| \frac{f(x) - P_n(x)}{(\sqrt{1 - x^2} + 1/n)^{r+\beta}} \right|^p dx \right)^{1/p} \leq \frac{C}{n^{r+\alpha}},$$

where the constant C does not depend on n or f.

The paper of Potapov also contains analogous results when the Lebesgue measure is changed by the Chebyshev one. Some other results were presented in [290]. The following result follows from the works of Lebed and Potapov.

Theorem 2.11.3 (Lebed-Potapov). *For $\alpha \in (0, 1)$ and a function f, one has $f \in A^{(r+\alpha)}$ if and only if for each $n \geq r$ there exists a polynomial $P_n \in \mathbb{P}_n$, such that*

$$\left(\int_{-1}^{1} \left| \frac{f(x) - P_n(x)}{(\sqrt{1 - x^2} + 1/n)^{r+\alpha}} \right|^p dx \right)^{1/p} \leq \frac{C}{n^{r+\alpha}},$$

where the constant C does not depend on n or f.

Taking into account (2.25), it was natural to look for weighted spaces. In this way some class of functions can be studied, but the original problems (characterization of classical Lipschitz spaces in terms of the best algebraic approximation or a characterization of a class of functions with a given rate for the best algebraic approximation in terms of the classical Lipschitz classes) was not solved. Since weighted approximation will not be discussed here in detail, we have included only a few remarks.

The characterization of the class $H_p^{(r+\alpha)}$ was also considered by Motornyi in 1971. He verified that the quantity

$$\lambda_n(f) = \inf_{P \in \mathbb{P}} \left\| \frac{f(x) - P(x)}{(\Delta_n(x))^\alpha} \right\|_{L_p}$$

are unbounded in the class $H_p^{(\alpha)}$ and established that classes $H_p^{(r+\alpha)}$ and $A_p^{(r+\alpha)}$ are different for $0 < \alpha < 1$ and coincide for $\alpha = 1$. He also characterized some functions, but not in terms of approximation by polynomials (see Theorem 11 of [255]) on the whole interval. Oswald [277] extended some of the results of Motornyi to the case of moduli of smoothness of higher order.

In 1978 DeVore [89] showed that we can not obtain a result similar to Theorem 2.7.7, if in (2.25) we replace the uniform norm by the $L_p[-1, 1]$ norm, $1 \leq p < \infty$. That is, by considering

$$F_n(f, r, \alpha)_p = \inf_{p \in \mathbb{P}_n} \left\| \frac{f(x) - p(x)}{\Delta_n^{r+\alpha}(x)} \right\|_p. \tag{2.50}$$

In fact DeVore showed (with an incomplete proof) that for $0 < \alpha < 1$ and $1 \le p < \infty$,

$$\omega_1(f, t)_p = \mathcal{O}(t^\alpha) \qquad \Longrightarrow \qquad F_n(f, 0, \alpha)_p = \mathcal{O}(\log n).$$

Moreover, for each $0 < \alpha < 1$ there exists $f \in L_p[-1, 1]$ such that $\omega_1(f, t)_p = \mathcal{O}(t^\alpha)$ and $F_n(f, 0, \alpha)_p \ge C \log n$ for infinitely many n. As we remarked above, Motornyi proved that these quantities are not bounded when f varies on the class $H_p^{(\alpha)}$.

In 1972 Golischek presented a detailed study of this kind of weighted approximation [143]. For $1 \le p \le \infty$ set

$$E_n^{(\lambda)}(f)_p = \inf_{p \in \mathbb{P}_n} \|(\max\{1/n, \sqrt{1-x^2}\})^{-\lambda}(f(x) - p(x)\|_p$$

and consider the following question: under what conditions are the statements

$$E_n^{(\lambda)}(f)_p = \mathcal{O}(n^{-\beta}) \tag{2.51}$$

and

$$\|(\max\{1/n, \sqrt{1-x^2}\})^{r-\lambda} P_n^{(r)}(x)\|_p = \mathcal{O}(n^{r-\beta}) \tag{2.52}$$

equivalent, where $r \in \mathbb{N}$ and β is a real number $0 < \beta < r$? The answer is different for $\lambda \le 0$ and $\lambda > 0$.

If $\lambda \le 0$, Golitschek proved that (2.51) and (2.52) are equivalent.

For $\lambda > 0$ the situation is more complicated: if $r > \max\{\beta, (\lambda + \beta)/2\}$, then (2.51) implies (2.52). Moreover, if we assume $E_n^{(\lambda)}(f)_p = 0$, then (2.52) implies (2.51). Golitschek also constructed a class of functions for which both assertions are equivalent.

The following theorem generalizes some results of Motornii in [254].

Theorem 2.11.4 (Shalashova, [337]). *Fix $k \in \mathbb{N}$ and $p \in [1, \infty)$. Suppose $f \in L_p[-1, 1]$ and $\omega_k(f, t)_p \le \Psi(t)$, where $\Psi(t)$ is some positive function satisfying the conditions:*

1) *$\Psi(t)$ does not decrease,*
2) *$\Psi(\lambda t) \le (\lambda + 1)^k \Psi(t)$ for $\lambda > 1$.*

Then for any integer $n > k$, one can find an algebraic polynomial P_n of degree not greater than $(4k + 2)n + k - 1$ such that

$$\left\| \frac{f(x) - p_n(x)}{\Psi(\Delta_n(x))} \right\|_p \le A_k [\log(n + 1)]^{1/p},$$

where A_k is a constant depending only on k.

Since A_k does not depend on p, we arrive at the uniform estimate of Brudnyi by letting p tend to ∞ in the last inequality (for a bounded f). If $r = k + 1$, and $\omega_1(f^{(k)}, t) \le Ct^\alpha$ ($0 < \alpha \le 1$), we obtain from the last result a theorem of Motornyi [254].

Some authors have studied the best approximation of particular classes of functions. For instance, Nasibov considered the approximation by algebraic polynomials of functions of the form

$$f(x) = \int_{-1}^{1} \psi\left(\frac{x-t}{2}\right) \varphi(t) dt \tag{2.53}$$

in the metric of $L_p[-1,1]$.

Theorem 2.11.5 (Nasibov, [266]). *Let $1 \le p, r < \infty$ and assume that $\psi \in L_r[-1,1]$ and $\varphi \in L_p[-1,1]$. If f is defined by (2.53), then*

$$E_n(f)_p \le 2^{2-1/r} \|\varphi\|_p E_n(\psi)_r.$$

Dynkin used a complex variable method (pseudo-analytical extension of functions) to obtain some results. For $s > 0$ and $1 \le p \le \infty$ he gave a characterization of functions satisfying

$$\left(\sum_{k=1}^{\infty} \frac{1}{n} E_n(f)_{p,s}^p\right)^{1/p} < \infty,$$

where

$$E_n(f)_{p,s} = \inf_{p \in \mathbb{P}_n} \left(\int_{-1}^{1} \left|\frac{f(x) - p(x)}{\Delta_n^s(x)}\right| dx\right)^{1/p}.$$

Recall that for $r \in \mathbb{N}$ and $1 < p < \infty$, $W_p^r[-1,1]$ is the class of functions such that $f^{(r-1)}$ is absolutely continuous and $f^{(r)} \in L_p[-1,1]$.

Oswald [277] considered the classes W_m of increasing functions ω such that $\omega(h) \le 2^m \omega(h/2)$ and $H_{p,m}^\omega$ of functions in L_p such that $\omega_m(f,t)_p \le C(f)\omega(t)$.

Theorem 2.11.6 (Oswald, [277]). *Fix $m \in \mathbb{N}$ and $p \in [1, \infty)$. For each $f \in L_p[a,b]$ and $n \ge m - 1$,*

$$E_n(f)_p \le C(m)\omega_m\left(f, \frac{b-a}{n+1}\right)_p.$$

From the inequality

$$\omega_{m+r}(f,t)_p \le t^r \omega_m(f^{(r)}, t)_p,$$

it follows that, for $f \in W_p^r[a,b]$,

$$E_n(f)_p \le C(m+r)\left(\frac{b-a}{n+1}\right)^r \omega_m\left(f^{(r)}, \frac{b-a}{n+1}\right)_p.$$

It can be used to characterize class $H_{p,m}^\omega$. Given $f \in L_p[a,b]$, $1 \le p < \infty$, and $m \in \mathbb{N}$, it is known that for any interval $[c,d]$, $[a,b] \subset (c,d)$, there exists an extension f^* of f to $[c,d]$ such that,

$$\omega_m(f^*, h)_p \le C\omega_m(f,h)_p.$$

If ω satisfies the condition

$$h^m \int_h^H \frac{\omega(t)}{t^{m+1}} dt \leq c\omega(t),$$

then the following conditions $f \in H_{p,m}^{\omega}$ and $E_n(f^*)_p \leq C\omega(1/n)$ are equivalent.

Theorem 2.11.7 (Dynkin, [104]). *Fix $1 < p < \infty$ and $r \in \mathbb{N}$. For a function $f : [-1, 1] \to \mathbb{R}$ one has $f \in W_p^r[-1, 1]$ if and only if, for each $k \in \mathbb{N}_0$, there exists $P_{2^k} \in \Pi_{2^k}$ such that*

$$\int_{-1}^1 \left(\sum_{k=0}^{\infty} \frac{|f(x) - P_{2^k}(x)|^2}{(\Delta_{2^k}(x))^{2r}} \right)^{p/2} dx < \infty.$$

It was Operstein, in 1995 [275], who stated the theory as completed as in the uniform norm. Let $\omega : \mathbb{R}^+ \to \mathbb{R}^+$ satisfy the condition $\omega(s+t) \leq M(\omega(s) + \omega(t))$ and set $\rho_k(x) = 2^{-k}\sqrt{1 - x^2} + 2^{-2k}$. We use the customary notation for the mixed norm

$$\|A_k(\cdot)\|_{l_p(L_p)} = \|\{\|A_k(\cdot)\|_{L_p}\}_k\|_{l_p}.$$

That is

$$\|A_k(\cdot)\|_{l_p(L_p)} = \left(\sum_{k=1}^{\infty} \int_{-1}^1 |A_k(x)|^p \, dx \right)^{1/p} = \|\{\|A_k(\cdot)\|_{L_p}\}_k\|_{l_p}.$$

Theorem 2.11.8 (Operstein, [275]). *Fix $p \in [1, \infty]$ and $r \in \mathbb{N}$. There exists a constant $C = C(p, r)$ such that, for each $f \in L_p[-1, 1]$ and $k \in \mathbb{N}_0$, there exists an algebraic polynomial $\{P_k\}$ of degree at most $2^k + r - 2$ such that*

$$\left\| \frac{f - P_k}{\omega(\rho_k)} \right\|_{l_p(L_p)} \leq C \left\| \frac{\omega_r(f, 2^{-k})}{\omega(2^{-k})} \right\|_{l_p}.$$

Brudnyi's Theorem 2.5.3 follows from this one by setting $\omega(t) = \omega_r(f, t)_p$ and $p = \infty$.

Theorem 2.11.9 ([275]). *Let f be a function defined on $[-1, 1]$. If there exists a sequence $\{P_k\}$ of algebraic polynomials of degree at most $2^k - 1$ such that*

$$\left\| \frac{f - P_k}{\omega(\rho_k)} \right\|_{l_p(L_p)} \leq 1,$$

then for every $r \in \mathbb{N}$,

$$\omega_r(f, t)_p \leq C t^r \left[\int_t^1 \left(\frac{\omega(u)}{u^r} \right)^q \frac{du}{u} \right]^{1/q}, \quad \frac{1}{p} + \frac{1}{q} = 1,$$

where the constant C depends only on r and p.

When $p = \infty$ we obtain the Timan inverse result. With these theorems one has the characterization of $\mathrm{Lip}(\alpha, p)$ spaces.

Theorem 2.11.10 (Operstein, [275]). *A function* $f : [-1, 1] \to \mathbb{R}$ *belongs to* $\mathrm{Lip}(\alpha, p)$ *if and only if there exists a sequence* $\{P_k\}$ *of algebraic polynomials of degree at most* 2^k ($k = 0, 1, \dots$) *such that*

$$\|(f - P_k) \min\{1, t/\rho_k\}^s\|_{l_p(L_p)} = \mathcal{O}(t^\alpha), \quad 0 < \alpha < s.$$

The idea of using $\min\{1, t/\rho_k\}$ for a characterization of $\mathrm{Lip}(\alpha, p)$ appears in [89], where it is proved that for each function $f \in \mathrm{Lip}(\alpha, p)$, $0 < \alpha < 1$, there exists a polynomial P_k such that $\|(f - P_k) \min\{1, t/\rho_k\}\|_{l_p(L_p)} = \mathcal{O}(t^\alpha)$. As we remarked above, Motornyi and DeVore showed that the direct analogue $(\|(f - P_n)\rho_n^{-\alpha}\}\|_{L_p} \leq C$ does not characterize $\mathrm{Lip}(\alpha, p)$ when $p < \infty$.

2.12 L_p, $0 < p < 1$

The behavior os the best approximation in L_p space, for $0 < p < 1$ is not the same as in the case $p \geq 1$. For instance, the difference $f(x) - P_n(x)$ (where P_n is a polynomial of the best approximation) must not oscillate at least at $n + 1$ point. For studies concerning this problem see [160], [402], [403], [404] and [405].

For $0 < p < 1$ the functional

$$\|f\|_p = \left(\int_a^b |f(x)|^p \, dx \right)^{1/p}$$

is not a norm, but the notation $\|f\|_p$ is used in this case for the sake of convenience.

Some smoothing processes which are usually applied in approximation theory do not work well in L_p spaces ($0 < p < 1$). Even more, the common definition of Sobolev spaces gives place to spaces with a trivial dual (see [279]). Thus, the ideas associated to K-functionals can not be used. There are also differences with the classical spaces related with the connection between smoothness and the existence of derivatives. In [214] Kortov studied this last topic.

In 1975 Storozhenko, Krotov and Oswald (Osval'd) presented direct and converse results for trigonometric approximation in the space of periodic functions $L_p[0, 2\pi]$, for $0 < p < 1$ [356]. The extension of the classical theory to this setting was motivated by some problems related with embedding theorems (see [352]). In [357] Storonzenko and Oswald presented estimates with the second-order modulus.

Theorem 2.12.1. *If* $0 < p < 1$ *and* $f \in L_p[0, 2\pi]$, *then*

$$E_n^*(f)_p \leq C_p \, \omega \left(f, \frac{\pi}{n+1} \right)_p,$$

where

$$E_n^*(f)_p = \inf_{T_n \in \mathbb{T}_n} \|f - T_n\|_p.$$

Moreover, for $n = 0, 1, \ldots,$

$$\omega\left(f, \frac{1}{n+1}\right)_p \le \frac{C_p}{n+1}\left(\sum_{j=0}^n (j+1)^{p-1}(E_j(f)_p)^p\right)^{1/p}. \tag{2.54}$$

Similar results were obtained in the same year by V.I. Ivanov [161], but he used moduli of smoothness of higher order:

$$\omega_k\left(f, \frac{1}{n}\right)_p \le \frac{C_{p,k}}{n^k}\left(\sum_{j=0}^n (j+1)^{kp-1}(E_j(f)_p)^p\right)^{1/p}.$$

In [356] and [161] a Bernstein inequality was proved for spaces $L_p[0, 2\pi]$, $0 < p < 1$, in the form

$$\|T^{(r)}\|_p \le C(p)n^r \|T\|_p.$$

Other proofs were given by Ivanov [162], Oswald [276], Nevai [269] and Runovskii [320]. The best result was presented by Arestov.

Theorem 2.12.2 (Arestov, [2]). *For $0 < p < 1$, $n, r \in \mathbb{N}$ and $T_n \in \mathbb{T}_n$ one has*

$$\|T_n^{(r)}\|_p \le n^r \|T_n\|_p.$$

In [320] and [321] Runovskii constructed some linear polynomial operators and obtained direct results in $L_p[0, 2\pi]$ ($0 < p < 1$) in the periodical case.

For $0 < p < \infty$ and $\mu \ge -1/p$, Khodak considered the spaces $L_{p,\mu}[-1,1]$ of functions f for which

$$\|f\|_{p,\mu} = \left(\int_{-1}^1 |f(x)(\sqrt{1-x^2})^\mu|^p\right)^{1/p} < \infty.$$

A function $f \in A_{p,\mu}^{\alpha,\beta}$ if

$$\left(\int_0^\pi \left|\frac{f(\cos(\gamma+t)) - f(\cos\gamma)}{(\sin\gamma + |\sin t|)^\beta}\right|^p (\sin\gamma)^{1+\mu p} d\gamma\right)^{1/p} \le C |\sin t|^\alpha.$$

A function $f \in \overline{A}_{p,\mu}^{\alpha,\beta}$ if

$$\left(\int_0^\pi \left|\frac{f(\cos(\gamma+t_1)) + f(\cos(\gamma+t_2)) - 2f(\cos(\gamma+(t_1+t_2)/2))}{(\sin\gamma + |\sin t|)^\beta}\right|^p (\sin\gamma)^{1+\mu p} d\gamma\right)^{1/p}$$
$$\le C|\sin t|^\alpha,$$

where $t = |t_1| + |t_2|$. Here β is a real number, for $p \ge 1$ we consider that $0 < \alpha \le 1$ and, for $0 < p < 1$, $0 < \alpha \le 1/p$.

For $p \geq 1$, and $t_1 = -t_2$ these spaces coincide with the one studied by Lebed and Potapov.

Theorem 2.12.3 (Khodak, [189]). *Let* $f \in L_{p,\mu}[-1,1]$, $0 < p < 1$, $\mu \geq -1/p$, $0 < \alpha < 2$,

$$-2/p - 2 - \mu + \alpha \ < \ -\beta \ < \ \alpha - 1/p - \mu.$$

In order that $f \in A_{p,\mu}^{\alpha,\beta}$ *for* $0 < \alpha < 1$ *or* $f \in \overline{A}_{p,\mu}^{\alpha,\beta}$ *for* $0 < \alpha < 2$, *it is necessary and sufficient that there exists a sequence* $\{P_n\}$, $P_n \in \mathbb{P}_n$ *such that*

$$\left\| [f(x) - P_n(x)] \left(\sqrt{1 - x^2} + \frac{1}{n} \right)^{-\beta} \right\|_{p,\mu} \leq \frac{C}{n^\alpha},$$

where the constant C *does not depend on* f *and* n.

As Ditzian showed we can not extend the results related with simultaneous approximation to the case $0 < p < 1$.

Theorem 2.12.4 (Ditzian, [97]). *For each* $0 < p < 1$ *there exists a function* $f \in A.C.[-1,1]$ *for which we can not find a sequence* $\{p_n\}$, $p_n \in \mathbb{P}_n$ *such that*

$$\|f - p_n\|_p \leq C\omega_2(f, 1/n)_p \qquad and \qquad \|f' - p_n'\|_p \leq C\omega(f', 1/n)_p.$$

The same assertion holds if we replace the usual moduli by the Ditzian-Totik one.

2.13   The Whitney theorem

Another form for the direct results in approximation by algebraic polynomials is due to Whitney.

Theorem 2.13.1 (Whitney, [407] and [408]). *For any* $n \in \mathbb{N}$ *there exists a constant* $W_\infty(n)$ *such that, for every bounded function* $f : [a,b] \to \mathbb{R}$ *there exists a polynomial* $P_{n-1}(f) \in \mathbb{P}_{n-1}$ *satisfying*

$$\|f - P_{n-1}(f)\| \leq W_\infty(n)\,\omega_n\left(f, \frac{b-a}{n}\right).$$

In fact, this was proved by Burkill in 1952 [43] for $n = 1, 2$, who also conjectured that the inequality holds for $n \geq 3$. In 1957 Whitney verified the conjecture for continuous functions and in 1959 for bounded functions. The proof of Whitney, as the one due to Burkill, used the polynomial P which interpolates f over a uniform net

$$P\left(\frac{k}{n-1}\right) = f\left(\frac{k}{n-1}\right), \qquad (k = 0, 1, \ldots, n - 1).$$

If we consider the polynomial $Q_{n-1}(f)$ which interpolates f at a uniform net of node, then we can also consider the inequalities

$$\|f - Q_{n-1}(f)\| \leq W'_\infty(n)\,\omega_n\left(f, \frac{b-a}{n}\right).$$

In 1964 Brudnyi found a new proof of the Whitney theorem. He used some smoothing of the function by means of linear combinations of Steklov-type functions. With the new method of proof he was able to extend the result to L_p spaces, with $1 \leq p < \infty$. In 1977 Storozhenko extended the Whitney theorem for algebraic approximation to $L_p[a,b]$ spaces, for $0 < p < 1$ (see also [358]).

Theorem 2.13.2 (Brudnyi, [39] and Storozhenko, [353]). *Suppose* $0 < p < \infty$, $f \in L_p(a,b)$ *and n is an arbitrary natural number, then*

$$E_{n-1}(f)_p \leq W_p(n)\,\omega_n\left(f, \frac{b-a}{n}\right)_p,$$

where $W_p(n)$ depends not on f.

Another proof was presented in [355] by Storozhenko and Kryakin.

In [354] Storozhenko presented the inequality: for $0 < p < 1$, $f \in L_p[-1,1]$, $k \in \mathbb{N}$ and $n \geq k - 1$,

$$E_n(f)_p \leq C_{p,k}\omega_k\left(f, \frac{1}{n+1}\right)_p.$$

A similar inequality appeared in [342] but only for the first modulus. Another proof was given by Khodak in [190].

The proof of Whitney can not be used as the estimate of the constants. Whitney proved that

$$\frac{1}{2} \leq W_\infty(n)$$

and found some bounds for some values of n. For instance

$$1 \leq W_\infty(1) \leq 2, \quad 1 \leq W_\infty(2) \leq 2.$$

The Whitney theorem has been studied by several Bulgarian mathematicians. In 1982 Sendov conjectured that $W_\infty(n) \leq 1$ [331].

This motivated several papers, shown in the table on top of the next page.

The inequality $W_\infty(n) \leq 1$ has been verified only for a few values of n: Whitney $n = 3$ [407], Kryakin $n = 4$ [219] and Zhelnov $k = 5, 6, 7, 8$, [416].

In [221] Kryakin and Takev proved that $W'_\infty(n) \leq 5\omega_n(f, 1/n)$.

Tunc [391] considered Whitney-type theorems in the form

$$E_{k+r+1}(f, [a,b])_\infty \leq W_\infty(k,r)\left(\frac{b-a}{k}\right)^r \omega_k\left(f, \frac{b-a}{k}, [a,b]\right)_\infty.$$

He found upper bounds for $W(k,2)$, $W(1,r)$ and $W(2,r)$.

Year	Author	Reference	Estimate
1952	Burkill	[43]	$W_\infty(2) = 1/2,$
1964	Brudnyi	[39]	$W_\infty(n) \le Cn^{2n},$
1985	Ivanov-Takev	[174]	$W_\infty(n) \le C(n \ln n),$
1985	Binev	[31]	$W_\infty(n) \le C\,n,$
1985	Sendov	[332]	$W_\infty(n) \le C,$
1986	Sendov	[333]	$W_\infty(n) \le 6,$
1985	Sendov-Takev	[335]	$W_1 \le 30,$
1989–90	Kryakin	[215], [216]	$W_\infty(n) \le 3,$
1989	Sendov-Popov	[334]	$W_\infty(n) \le 3,$
1990	Kryakin	[216]	$W_p(n) \le 11,$
1992	Kryakin-Kovalenko	[220]	$W_1 \le 6.4,$
1992	Kryakin-Kovalenko	[220]	$W_p \le 9,$
1995	Kryakin	[217], [218]	$W_\infty(n) \le 2.$
2002	Gilewicz-Kryakin-Shevchuk	[142]	$W_\infty(n) \le 2 + e^{-2}.$

2.14   Other classes of functions

Bernstein [30] characterized $C^\infty[a, b]$ as follows: $f \in C^\infty[a, b]$ if and only if each $k \in \mathbb{N}$,

$$\lim_{n \to \infty} n^k E_n(f) = 0.$$

Some subclasses of functions of $C^\infty[a, b]$ has been studied by Brudnyi-Gopengauz [41] Babenko [4] and Motornyi [256].

Chapter 3

Looking for New Moduli

Different authors have tried to used other forms of measuring the smoothness of functions. In the first section of this chapter we present some of the ideas associated to the works of Potapov. In the second section we analyze the circle of ideas developed by Butzer and his collaborators.

3.1 The works of Potapov

Potapov began to consider the approximation by algebraic polynomials in L_p spaces in 1956 [287], where he follows Timan's ideas. In 1960 and 1961 he obtained results in which the usual translation was modified ([289] and [290]).

Let $L_{p,\alpha,\beta}[-1,1]$ be the space of all functions f for which

$$\|f\|_{p,\alpha,\beta} = \|f(x)(1-x)^\alpha(1+x)^\beta\|_p < \infty$$

and $E_n(f)_{p,\alpha,\beta}$ be the best approximation by algebraic polynomials in this space. That is

$$E_{n,\alpha,\beta}(f) = \inf_{P_n \in \mathbb{P}_n} \|f - P_n\|_{p,\alpha,\beta}.$$

When $\alpha = \beta$ we simply write $L_{p,\alpha}[-1,1]$ and $E_n(f)_{p,\alpha}$.

Let us denote by $\omega(f,t)_{p,\alpha,\beta}$ the usual modulus of continuity of f in the metric of $L_{p,\alpha,\beta}[-1,1]$. That is

$$\omega(f,t)_{p,\alpha,\beta} = \sup_{|h| \leq t} \|f(x+h) - f(x)\|_{p,\alpha,\beta},$$

with the usual restriction relative to the interval ($f(x+h) - f(x) = 0$, if $x+h > 1$). In 2000 Potapov proved that in $L_{p,\alpha,\beta}[-1,1]$ the usual Lipschitz classes can not be characterized by the best approximation in the same form as in the case of trigonometric approximation [300].

Let us set

$$G(f, x, t) = \frac{1}{2}[f(x\cos t + \sqrt{1 - x^2}\sin t) + f(x\cos t - \sqrt{1 - x^2}\sin t)] \qquad (3.1)$$

and define

$$\widetilde{\omega}(f, t)_{p,\alpha,\beta} = \sup_{|h| \le t} \|f(x) - G(f, x, h)\|_{p,\alpha,\beta}.$$

Theorem 3.1.1 (Potapov, [289]). *Fix $p \in [1, \infty]$, $\alpha = \beta = -1/(2p)$ and $\gamma \in (0, 1)$. For $f \in L_{p,\alpha,\beta}[-1, 1]$ the following assertions are equivalent:*

(i) *There exists a constant M such that $\widetilde{\omega}(f, t)_{p,\alpha,\beta} \le Mt^\gamma$.*
(ii) *There exists a constant K such that, for all $n \in \mathbb{N}$, $E_n(f)_{p,\alpha,\beta} \le K/n^\gamma$.*

For the un-weighted case ($\alpha = \beta = 0$) and $p = 2$, Zhidkov obtained another characterization. Define

$$\widehat{\omega}(f, t)_{p,\alpha,\beta} = \sup_{|h| \le t} \|f(x) - H(f, x, h)\|_{p,\alpha,\beta},$$

where

$$H(f, x, h) = \frac{1}{\pi} \int_{-1}^{1} f(x\cos t + y\sin t\sqrt{1 - x^2}) \frac{dy}{\sqrt{1 - y^2}}. \qquad (3.2)$$

Theorem 3.1.2 (Zhidkov, [417]). *If $\gamma \in (0, 1)$, for a function $f \in L_2[-1, 1]$ there exists a constant M such that $\widehat{\omega}(f, t)_{2,0,0} \le Mt^\gamma$ if and only if there exists a constant K such that, for all $n \in \mathbb{N}$, $E_n(f)_{2,0,0} \le K/n^\gamma$.*

Let us recall another result of Zhidkov.

Theorem 3.1.3 ([417]). *For $f \in L_2[-1, 1]$ one has $E_n(f)_2 \le C/n^{s+\gamma}$ if and only if*

$$\left(\int_{-1}^{1} \left(\frac{d^s f_h(x)}{dx^s} - \frac{d^s f(x)}{dx^s} \right)^2 (1 - x^2)^s dx \right)^{1/2} \le C h^\gamma,$$

where $h > 0$, $n > s$, $0 < \gamma < 1$ and

$$f_h(x) = \frac{1}{\pi} \int_{0}^{\pi} f(x\cos h + \sqrt{1 - x^2}\sin h\cos t)dt.$$

In [292] Potapov considered the problem of characterizing all functions $f \in L_{p,\alpha,\beta}[-1, 1]$ for which there exists a sequence of algebraic polynomials satisfying

$$\left\| (f(x) - P_n(x)) \left(1 - x + \frac{1}{n^2}\right)^{\rho_1} \left(1 + x + \frac{1}{n^2}\right)^{\rho_2} \right\|_{p,\alpha,\beta} \le C\frac{1}{n^{r+\gamma}}.$$

The case $\alpha = \beta \ge -1/(2p)$ and $\rho_1 = \rho_2$ has been studied previously by him in [291]. The results are given in terms of a generalized translation.

For $f : [-1, 1] \to \mathbb{R}$, consider it a Fourier-Jacobi series

$$f(x) \sim \sum_{k=0}^{\infty} a_k P_k^{(\alpha,\beta)}(x)$$

where $\{P_k^{(\alpha,\beta)}\})$ is the sequence of Jacobi polynomials. That is, they are the orthogonal polynomials in $[-1, 1]$ with respect to weight $(1 - x)^{\alpha}(1 + x)^{\beta}$, with the normalization $P_k^{(\alpha,\beta)}(1) = 1$.

Let us consider the associated series

$$T_h(f, x, \alpha, \beta) = \sum_{k=0}^{\infty} a_k P_k^{(\alpha,\beta)}(x) P_k^{(\alpha,\beta)}(h). \tag{3.3}$$

Assume that, for each $h \in [-1, 1]$, there exists a function g_h such that (3.3) holds if the Fourier-Jacobi series of g_h holds, then we consider that (3.3) is the Fourier-Jacobi series of f with generalized translation $x + h$. The functions were called the *generalized translation* by Löfström and Peetre in [238].

Now the generalized modulus is defined by

$$\omega(f, t, \alpha, \beta)_{p,\alpha,\beta} = \sup_{|s| \le t} \|f(x) - T_{\cos s}(f, x, \alpha, \beta)\|_{p,\alpha,\beta}. \tag{3.4}$$

It is not a simple task to find a simple expression for the generalized translation $T_h(f, x, \alpha, \beta)$.

In the case $\alpha = \beta = -1/2$, the Fourier-Jacobi polynomials are just the Chebyshev polynomials: $P_n^{(-1/2,-1/2)}(x) = T_n(x) = \cos(n \arccos x)$. It can be proved that

$$T_{\cos t}(f, x, -1/2, -1/2) = G(f, x, t),$$

where $G(f, x, t)$ is the function defined (3.1). In this case the direct and converse results were recalled in Theorem 3.1.1.

In the case $\alpha = \beta = 0$, the Fourier-Jacobi polynomials are the Legendre polynomials and the translation has the form

$$T_{\cos t}(f, x, 0, 0) = H(f, x, t),$$

where $H(f, x, t)$ is defined by (3.2). In this case the direct and converse results were recalled in Theorem 3.1.2 (in L_2 spaces).

Recall that the Legendre polynomial $P_n(x)$ of degree n is defined by

$$P_n(x) = \frac{(-1)^n}{2^n n!} \frac{d^n}{dx^n} (1 - x^2)^n, \qquad (x \in [-1, 1], \quad n \in \mathbb{N}_0). \tag{3.5}$$

For $\alpha = \beta > -1/2$, the Fourier-Jacobi polynomials are the Gegenbauer polynomials and the translation has the form

$$T_{\cos t}(f, x, \alpha, \alpha) = \frac{1}{\gamma(\alpha)} \int_{-1}^{1} f(x \cos t + y\sqrt{1 - x^2} \sin t)(1 - y^2)^{\alpha - 1/2} dy, \quad (3.6)$$

where $\gamma(\alpha) = \int_{-1}^{1}(1 - y^2)^{\alpha - 1/2} dy$. In this case the direct and converse results were given by Rafalson [311] and Pawelke [278]. Rafalson extended the theorem of Zhidkov [417] for the case $\alpha > 0$. Zhidkov and Rafalson only considered approximation in L_2 spaces.

For $\alpha > \beta = -1/2$, the translation is given by

$$T_{\cos t}(f, x, \alpha, \beta) = \frac{1}{\gamma_1(\alpha, \beta)} \int_{-1}^{1} f(\Phi_1(x, t, y))\Theta_1(y) dy, \quad (3.7)$$

where

$$\Phi_1(x, t, y) = x \cos t + y\sqrt{1 - x^2} \sin t - (1 - y^2)(1 - x)\sin^2(t/2),$$
$$\Theta_1(y) = (1 - y^2)^{\alpha - 1/2}.$$

and $\gamma_1(\alpha, \beta)$ is chosen from the condition $T_{\cos t}(1, x, \alpha, \beta) = 1$.

For $\alpha > \beta > -1/2$, the Fourier-Jacobi polynomials are the Jacobi polynomials and the translation has the form

$$T_{\cos t}(f, x, \alpha, \beta) = \frac{1}{\gamma_2(\alpha, \beta)} \int_{0}^{1} \int_{-1}^{1} f(\Phi_2(x, t, r, y))\Theta_2(r, y) dy dr, \quad (3.8)$$

where

$$\Phi_2(x, t, r, y) = x \cos t + ry\sqrt{1 - x^2} \sin t - (1 - r^2)(1 - x)\sin^2(t/2),$$
$$\Theta_2(r, y) = (1 - r^2)^{\alpha - \beta - 1} r^{2\beta + 1}(1 - y^2)^{\alpha - 1/2}$$

and $\gamma_2(\alpha, \beta)$ is chosen from the condition $T_{\cos t}(1, x, \alpha, \beta) = 1$. In this case the direct and converse results were given by Potapov in [292].

The case $\alpha = 0$ and $\beta > -1$, was studied by Potapov, Fedorov and Fraguela in [309] and [308]. They wrote the generalized translation as

$$T_t^{\beta}(f, x) = \frac{1}{\pi \cos^{2\beta}(t/2)} \int_{0}^{\pi} f(\cos s)\left(\frac{1 + \cos s}{1 + x}\right)^{\beta} \cos(2\beta r) du$$

where

$$|t| < \pi, \quad \cos s = x \cos t + \cos u \sin t \sqrt{1 - x^2}, \quad 0 \le r \le \pi$$

and

$$\cos r = \frac{\sqrt{1 + x} \cos(t/2) + \cos u\sqrt{1 - x} \sin(t/2)}{\sqrt{1 + x \cos t + \cos u\sqrt{1 - x^2} \sin t}}.$$

With this translation the modulus is defined by

$$\Omega(f,\delta)_{p,\beta} = \sup_{|t|\leq\delta} \|[T_t^\beta(f,x) - f(x)](1+x)^\beta\|_p.$$

Theorem 3.1.4 (Potapov-Fedorov, [308]). *Suppose that $\beta > -1/2$ and $1 \leq p \leq \infty$. There exists a positive constant C_1 and C_2 such that, for every $f \in L_{p,0,\beta}[-1,1]$ and $n \in \mathbb{N}$,*

$$C_1 E_n(f)_{p,0,\beta} \leq \Omega\left(f,\frac{1}{n}\right)_{p,\beta} \leq \frac{C_2}{n^2} \sum_{k=1}^n k E_k(f)_{p,0,\beta}.$$

For given ν and μ, assume that the translation $H(f,t,\nu,\mu)$ is defined by (3.1) when $\nu = \mu = -1/2$, by (3.6) when $\nu = \mu > -1/2$, by (3.7) when $\nu > \mu = -1/2$ and by (3.8) when $\nu > \mu > -1/2$. With this selection define the modulus

$$\omega(f,\delta,\mu,\nu)_{p,\alpha,\beta} = \sup_{|t|\leq\delta} \|f(x) - H(f,t,\nu,\mu)\|_{p,\alpha,\beta}.$$

Theorem 3.1.5 (Potapov, [297]). *Fix $p \in [1,\infty]$ and $\alpha \geq \beta \geq -1/(2p)$. Assume that ν and μ are chosen following the rules:*

$$\begin{array}{lll}
\mu = \nu = -1/2, & \text{if} & \alpha = \beta = -1/(2p), \\
\mu = -1/2, \nu > \alpha - 1/2 + 1/(2p), & \text{if} & \alpha > \beta = -1/(2p), \\
\nu = \mu > \alpha - 1/2 + 1/(2p), & \text{if} & \alpha = \beta > -1/(2p), \\
\mu > \beta - 1/2 + 1/(2p), & & \\
\nu > \mu + \alpha - \beta, & \text{if} & \alpha > \beta > -1/(2p).
\end{array}$$

There exist positive constants C_1 and C_2 such that, for all $f \in L_{p,\alpha,\beta}$,

$$C_1 E_n(f)_{p,\alpha,\beta} \leq \omega\left(f,\frac{1}{n},\nu,\mu\right)_{p,\alpha,\beta} \leq \frac{C_2}{n}\sum_{k=1}^n k E_k(f)_{p,\alpha,\beta}.$$

Some other results concerning Jacobi weights were given by Potapov in [296]. Some extensions to moduli of higher order were presented by Tankaeva in [370] and by Potatov and Kazimirov in [310]. Other results for Jacobi weights were obtained by Potapov, Berisha and Berisha in [307].

In 1999 Potapov provided another modulus using a non-symmetric generalized operator of translation. He considered the expression given in (3.3) as the symmetrical case and replace the term $P_k^{(\alpha,\beta)}(h)$ by $\varphi_k(h)$. The new formula

$$T_h(f,x,\alpha,\beta) = \sum_{k=0}^\infty a_k P_k^{(\alpha,\beta)}(x)\varphi_k(h), \qquad (3.9)$$

is called non symmetric.

For the case $\alpha = \beta$, he considered the translation

$$T_t(f, x) = \frac{1}{\pi(1 - x^2)} \int_0^\pi f(\Phi_3(x, t, s))\Theta_3(x, t, s)ds,$$

where

$$\Phi_3(x, t, s) = x \cos t + \cos s \sin t \sqrt{1 - x^2}$$
$$\Theta_3(x, t, s) = 1 - (\Phi_3(x, t, s))^2 - 2\sin^2 t \sin^2 s + 4(1 - x^2)\sin^2 t \sin^4 s$$

and the modulus

$$\widetilde{\omega}(f, \delta)_{p,\alpha} = \sup_{|t| \leq \delta} \|T_t(f, x) - f(x)\|_{p,\alpha}.$$

Theorem 3.1.6 (Potapov, [298]). *Fix p, α and r such that $r \in (0, 2)$ and*

$$
\begin{array}{lll}
\alpha \in (1/2, 1], & if & p = 1, \\
\alpha \in (1 - 1/(2p), 3/2 - 1/(2p)), & if & 1 < p < \infty, \\
\alpha \in [1, 3/2), & if & p = \infty.
\end{array}
$$

For a function $f \in L_{p,\alpha}[-1, 1]$ the following assertions are equivalent:

(i) $\qquad\qquad\qquad\qquad E_n(f)_{p,\alpha} \leq C(f)\, n^{-r},$

(ii) $\qquad\qquad\qquad\qquad \widetilde{\omega}(f, \delta)_{p,\alpha} \leq C(f)\delta^r.$

Extension to moduli of smoothness of order r were given by Potapov and Berisha in [304] (see also [305]).

In the case $\alpha = \beta + 1$, the translation is defined by

$$T_y(f, x) = \frac{4}{\pi} \int_{-1}^1 f(R)\psi(x, y, z)\frac{dz}{\sqrt{1 - z^2}},$$

where

$$\psi(x, y, z) = \frac{\cos(u + \mu - u_1)(1 - R)\sqrt{1 - R^2}}{(1 + y)^2(1 - x)\sqrt{1 - x^2}},$$

$$R = xy + z\sqrt{1 - x^2}\sqrt{1 - y^2},$$

$$\cos u_1 = z, \qquad \sin u_1 = \sqrt{1 - z^2},$$

$$\cos u = \frac{-\sqrt{1 - y^2}x + yz\sqrt{1 - x^2}}{\sqrt{1 - R^2}}, \qquad \sin u = \frac{\sqrt{1 - x^2}\sqrt{1 - z^2}}{\sqrt{1 - R^2}},$$

$$\cos \mu = \frac{z(1 - xy) - \sqrt{1 - x^2}\sqrt{1 - y^2}}{1 - R}, \qquad \sin \mu = \frac{\sqrt{1 - z^2}(y - x)}{1 - R}.$$

Now define

$$\widehat{\omega}(f, \delta)_{p,\alpha,\beta} = \sup_{|t| \leq \delta} \|T_{\cos t}(f, x) - f(x)\|_{p,\alpha,\beta}.$$

Theorem 3.1.7 (Potapov, [299]). *Fix $p \in [1, \infty]$ and assume*

$$
\begin{aligned}
&\alpha \in (0, 1/2], && p = 1, \\
&\alpha \in (1/2 - 1/(2p), 1 - 1/(2p)), && 1 < p < \infty, \\
&\alpha \in [1/2, 1), && p = \infty.
\end{aligned}
$$

There exist positive constants C_1 and C_2 such that, for all $f \in L_{p,\alpha+1,\alpha}$,

$$
C_1 E_n(f)_{p,\,\alpha+1,\,\alpha} \leq \widehat{\omega}(f, 1/n)_{p,\,\alpha+1,\,\alpha} \leq \frac{C_2}{n^2} \sum_{k=1}^{n} k E_k(f)_{p,\,\alpha+1,\,\alpha}.
$$

3.2 Butzer and the method of Fourier transforms

The methods of Fourier transforms can be used to prove of the classical assertions for trigonometric approximation into the Jacobi-weighted frame. Several authors are related to this topic. Ganser [130] introduced the modulus of continuity in the Jacobi frame.

The works began with Bavinck ([19] and [20]) and Scherer and Wagner [330] in 1972. Butzer and Stens ([55], [56] and [57]) introduced the Chebyshev transform method and Butzer, Stens and Wehrens ([58], [59], [60] and [350]) the Legendre transform method. Finally, the Jacobi transform method was presented in [60]. Some results concerning Gegenbauer-weights were given by Löfström [237].

One of the disadvantages of the Jacobi transform method is that derivatives and Lipschitz classes are defined in terms of a generalized translation. Let us present some ideas taken from [55].

As usual, $C[-1, 1]$ denotes the set of all continuous real-valued functions f defined on $[-1, 1]$ with the sup norm. Let L_w^p, $1 \leq p < \infty$, be the set of all measurable real-valued f on $[-1, 1]$ for which the norm

$$
\|f\|_p = \left(\frac{1}{\pi} \int_{-1}^{1} |f(u)|^p\, w(u) du \right)^{1/p}
$$

$w(x) = 1/\sqrt{1 - x^2}$, is finite.

Below, X stands for one of the Banach spaces $C[-1, 1]$ or L_w^p.

For $f \in X$, the kth Chebyshev-Fourier coefficient is defined by

$$
\mathbb{T}[f](k) = [f]^\wedge(k) = \frac{1}{\pi} \int_{-1}^{1} f(u) T_k(u) \frac{du}{\sqrt{1 - u^2}} \tag{3.10}
$$

where $T_k(x) = \cos(k \arccos x)$ $(x \in [-1, 1])$, is a Chebyshev polynomial of degree k.

The classical translation of a function $f(x)$ by h, namely $f(x + h)$, is replaced by

$$
(\tau_h f)(x) = \frac{f\left(xh + \sqrt{(1 - x^2)(1 - h^2)}\right) + f\left(xh - \sqrt{(1 - x^2)(1 - h^2)}\right)}{2} \tag{3.11}
$$

$(x, h \in [-1, 1])$. This translation has the advantage that it is an operator from X into itself and satisfies $\lim_{h \to 1^-} \|\tau_h f - f\|_X = 0$. Thus one can define the *Chebyshev derivative* as the function $g \in X$ for which

$$\lim_{h \to 1^-} \left\| \frac{f - \tau_h f}{1 - h} - g \right\|_X = 0,$$

whenever such a function exists, and then we write $D^1 f = g$. Derivatives D^r of higher order $r = 2, 3, \ldots$ are defined iteratively.

The set of all $f \in X$ for which $D^r f$ exists is denoted by W_X^r. It was proved in [56] that, for $f \in X$, one has $f \in W_X^r$ if and only if there exists $g \in X$ such that

$$(-k^2)^r \mathbb{T}[f](k) = \mathbb{T}[g](k). \tag{3.12}$$

In this case $D^r f = g - g^\wedge(0)$. If we define the convolution product of $f \in L_w^1$ and $g \in X$ by

$$(f * g)(x) = \frac{1}{2} \int_{-1}^{1} (\tau_x f)(u) g(u) w(u) du, \tag{3.13}$$

then $f * g \in X$, and its Chebyshev transform, satisfies

$$\mathbb{T}[f * g](k) = \mathbb{T}[f](k) \mathbb{T}[g](k). \tag{3.14}$$

The (right) difference of $f \in X$ of order $r \in \mathbb{N}$ with respect to the increment $h \in [-1, 1]$ is defined by

$$(\overline{\Delta}_h^1 f)(x) = (\tau_h f)(x) - f(x),$$
$$(\overline{\Delta}_h^r f)(x) = (\overline{\Delta}_h^1 (\overline{\Delta}_h^{r-1} f))(x).$$

With the notions presented above, the modulus of continuity and Lipschitz class are introduced as follows:

$$\omega_r^T(f, t) = \sup_{t \le h \le 1} \|(\overline{\Delta}_h^r f)\|_X, \qquad (t \in [-1, 1]) \tag{3.15}$$

and

$$\text{Lip}_r^T(\alpha, X) = \{ f \in X : \omega_r^T(f, t) = \mathcal{O}((1 - t)^\alpha) \}.$$

There are relations between these notions and the usual moduli of continuity. If $X = L_w^p$, we denote by $X_{2\pi}$ the L_p space of 2π-periodic functions.

Proposition 3.2.1. *For $f \in X$, $F \in X_{2\pi}$, $\eta \in [-1, 1]$, $\delta > 0$, $\alpha > 0$ and $r \in \mathbb{N}$, one has*

(i) $\omega_r^T(f, \eta) = \omega_{2r}(f \circ \cos, \arccos \eta)$,
(ii) *f belongs to $\text{Lip}_r^T(X; \alpha)$ if and only if $f \circ \cos$ belongs to $\text{Lip}_{2r}(X_{2\pi}; 2\alpha)$,*
(iii) *if F is even, then $F \in \text{Lip}_{2r}(X_{2\pi}; 2\alpha)$ if and only if $F \circ \arccos \in \text{Lip}_r^T(X; \alpha)$.*

In this setting Butzer and Stens presented an analogue of Theorem 1.2.1 for the best algebraic approximation.

Theorem 3.2.2. *Fix $r, r_1, r_2 \in \mathbb{N}$ and $0 < \alpha < 1$. For a function $f \in X$ the following assertions are equivalent:*

(i) $E_n(f; X) = \mathcal{O}(n^{-2(r+\alpha)})$, $(n \to \infty)$,

(ii) $\omega_1^L(D^r f; \delta) = \mathcal{O}((1-\delta)^\alpha)$, $(\delta \to 1-)$,

(iii) $\|D^{r_1} p_n^*(f)\|_X = \mathcal{O}(n^{-2(r+\alpha-r_1)})$, $(r_1 > r + \alpha, n \to \infty)$,

(iv) $f \in W_W^{r_2}$, $\|D^{r_2} f - D^{r_2} p_n^*(f)\|_X = \mathcal{O}(n^{-2(r+\alpha-r_2)})$, $(r_2 < r + \alpha, n \to \infty)$.

Proof. The method of proof follows the general approach of Theorem 1.4.1. Take $M_n = \mathbb{P}_n$, $\rho = r_1$, $\sigma = r_1$ and $s = r + \alpha$. Moreover, set $Y = W^{r_1}$, with seminorm $|\, g\, |_Y = \|D^{r_1} g\|_X$, and $Z = W^{r_2}$, with seminorm $|\, h\, |_Z = \|D^{r_2} h\|_X$. It can be proved that D^{r_2} is a closed operator (see [56], Corollary 4). Hence, in view of the closed graph theorem, Z becomes a Banach space under the norm $\|h\|_Z = \|h\|_X + \|D^{r_2} h\|_X$.

It is known that, for each $m \in \mathbb{N}$, there exist positive constants $D_1 = D_1(m)$ and $D_2 = D_2(m)$ such that, for $f \in X$ and $t \in (0, \pi]$

$$D_1 \omega_m^T(f, \cos t) \leq K(f, t^{2m}, X, W_X^m) \leq D_2 \omega_m^T(f, \cos t).$$

Taking into account that, for all $n \in \mathbb{N}$, $M_n \subset Y \subset X$, $M_n \subset Z \subset X$ and that, for $f \in X$ an element of the best approximation always exists in M_n, we only need to verify the Jackson- and Bernstein-type inequalities given in (1.17) and (1.18) hold.

The Bernstein-type inequality follows from the classical Bernstein inequality for trigonometric polynomials. In fact, if $n, m \in \mathbb{N}$ and $P_n \in \mathbb{P}_n$, then

$$\|D^m P_n\|_X = \|(P_n \circ \cos)^{(2m)} \circ \arccos\|_X \leq (2n)^{2m} \|P_n \circ \cos\|_{X, 2\pi},$$

where the last norm in computed on the interval $[0, 2\pi]$.

Now, let us consider the Jackson-type inequality. Let K_n be the Fejér-Korovkin operator (the formal definition is given in the section devoted to integral operators). Set $K_n^0 = I$, $K_n^1 = K_n$ and, $K_n^j = K_n^1(K_n^{j-1})$ $(j \in \mathbb{N})$. It can be proved that, for each $j \in \mathbb{N}$ and $f \in X$, $K_n^j(f) \in \mathbb{P}_n$.

Set

$$U_{r,n} = \sum_{j=1}^{r} (-1)^{j+1} \binom{r}{j} K_n^j.$$

Since

$$U_{r,n}(f) - f = (-1)^{r-1} (K_n - I)^r (f),$$

then (see Corollary 5.5.7 below)

$$\|U_{r,n}(f) - f\|_X = \|K_n((K_n - I)^{r-1}(f)) - (K_n - I)^{r-1}(f)\|_X$$

$$\leq \left(1 + \frac{\pi}{\sqrt{2}}\right)^2 \omega_1^T\left((K_n - I)^{r-1}(f), \cos\left(\sqrt{1 - \cos\frac{\pi}{n+2}}\right)\right).$$

In particular, there exits a constant C_1 such that, if $f \in W_X^r$, then

$$\|U_{r,n}(f) - f\|_X \leq \frac{C_1}{n^2}\|D^1((K_n - I)^{r-1}(f))\|_X.$$

This yields the Jackson-type inequality in the case $r = 1$.

If $r > 1$, taking into account that

$$D^1((K_n - I)^{r-1}(f)) = (K_n - I)^{r-1}(D^1(f))$$

and using the arguments given above we obtain a constant C_2 such that

$$\|U_{r,n}(f) - f\|_X \leq \frac{C_1}{n^2}\|(K_n - I)^{r-1}(D^1(f))\|_X \leq \frac{C_2}{n^4}\|D^2((K_n - I)^{r-2}(f))\|_X.$$

This yields the Jackson-type inequality in the case $r = 2$.

By repeating this process we obtain the general assertion. $\square$

If we compare these results with the ones we presented above, we notice several facts. Fuksman worked with the classical derivative concept and did not include assertions concerning higher-order moduli. Asadov [3] and Khalilova [187] followed a similar approach, however only for functions which are quadratically integrable with respect to a weight. Dzafarov [105] considered continuous functions and used a different notion of derivative. Finally, Bavinck [19] examined spaces with weight $(1 - x^2)^\beta(1 - x^2)^\gamma$ for certain values of β and γ, but he did not characterize the assertion $E_n(f) = \mathcal{O}(n^{-2r})$, $r \in \mathbb{N}$.

Other results related with the work of Butzer will be presented in the section devoted to Ditzian and Totik.

3.3   The τ modulus of Ivanov

In order to obtain characterizations for the second interpretation after (1.11), Ivanov used the τ modulus ([163] and [166]).

Given an arbitrary positive function δ and a non-negative continuous function w, define

$$\tau_k(f, w, \delta)_{r,p,[a,b]} = \|w(\cdot)\omega_k(f, \cdot, \delta(\cdot))_r\|_{p,[a,b]}$$

where

$$\omega_k(f, x, \delta(x))_r = \left(\frac{1}{2\delta(x)}\int_{-\delta(x)}^{\delta(x)}\left|\Delta_v^k f(x)\right|^r dv\right)^{1/r}.$$

In order to simplify we omit the index $[a, b]$. Moreover, when $w \equiv 1$ we omit w in the notation.

Another τ modulus is defined by

$$\tau_k(f,t)_p^* = \|\omega_k(f,\cdot,t)_r\|_{p,[a,b]}$$

where

$$\omega_k(f,x,t) = \sup\left\{|\Delta_h^k f(x)| : t, t+kh \in \left[x - \frac{kt}{2}, x + \frac{kt}{2}\right] \cap [a,b]\right\}.$$

Several properties of these moduli, as well as its connection with $\omega_k(f,t)_p$ and $\tau_k^*(f,t)_p$, were given in [163] and proved in [166].

Theorem 3.3.1. *If* $1 \le r, p$, $f, g \in L_{\max\{r,p\}}[a,b]$ *and* $\alpha \in \mathbb{R}$, *then*

(i) $\tau_k(f + g, w, \delta)r, p \le \tau_k(f, w, \delta)r, p + \tau_k(g, w, \delta)r, p,$

(ii) $\tau_k(\alpha f, w, \delta)r, p = |\alpha| \tau_k(f, w, \delta)r, p,$

(iii) $\tau_k(f, w_1, \delta)r, p \le \tau_k(f, w_2, \delta)r, p$, $0 \le w_1 \le w_2,$

(iv) $\tau_k(f, w, \delta)r, p_1 \le (b-a)^{1/p_1 - 1/p_2} \tau_k(f, w, \delta)r, p_2$, $1 \le p_1 \le p_2,$

(v) $\tau_k(f, w, \delta)r_1, p \le \tau_k(f, w, \delta)r_2, p$, $1 \le r_1 \le r_2.$

Theorem 3.3.2. *For* $p, r, s \ge 1$, $d > 0$, $n \in \mathbb{N}$ *and* $\alpha \ge 1$,

(i) $\tau_1(f, n\,d)_{1,p} \le n\,\tau_1(f, d)_{1,p},$

(ii) $\tau_1(f, \alpha\,d)_{1,p} \le (3 + [\alpha])\,\tau_1(f, d)_{1,p},$

(iii) $\tau_k(f, d)r, p \le C(k)\,\tau_{k-1}(f', d, d)s, p$, , $k \ge 2$, $f' \in L_{\max\{p,s\}},$

(iv) $\tau_k(f, d)r, p \le \omega_k(f, d)_p \le C(k)\tau_k(f, d)r, p$, $r \in [1, p],$

(v) $\tau_1(f, d)\infty, p \le \tau_1^*(f, d)_p \le 2\,\tau_1(f, d)\infty, p$, $f \in L_\infty,$

(vi) $\tau_k(f, d)\infty, p \le C(k)\,\tau_1^*(f, d)_p$, $f \in L_\infty$, $k \ge 2.$

Theorem 3.3.3. *Suppose that the weight* w *satisfies the following condition: for every* $x, t \in [-1, 1]$ *for which* $|x - t| \le \lambda(d\sqrt{1 - x^2} + d^2)$,

$$w(x) \le C(\lambda)\,w(t). \tag{3.16}$$

For $1 \le p, r, s \le \infty$, $d \le 1$ *and* $f \in L_{\max\{p,r\}}$ *(or* $f' \in L_{\max\{p,r\}}$, *or* $f^{(k)} \in L_p$*), then*

(i) $\tau_k(f, w, \Delta(d))r, p \le C(k)\,\|wf\|_p$, $r \le p,$

(ii) $\tau_k(f, w, \Delta(d))s, p \le C(k)\,\tau_{k-1}(f', w\Delta(d), \Delta((4k+2)d)r, p$, $k \ge 2,$

(iii) $\tau_k(f, w, \Delta(d))r, p \le C(k)\,\|\Delta^k(d)f^{(k)}\|_p$, $k \ge 1,$

(iv) $\tau_1(f, w, A\Delta(d))r, p \le C(A)\tau_1(f, w, \Delta(d))r, p$, $d \le (2A)^{-1}$, $A \ge 1,$

(v) $\tau_k(f, w, \Delta(d))r, p \le \tau_k(f, w, \Delta(d))s, p \le C(k)\,\tau_k(f, w, \Delta(d))r, p$, $1 \le r \le s \le p.$

The weight $w(x) = (d\sqrt{1 - x^2} + d^2)^\mu$ (μ real) satisfies (3.16) with a constant $C(\lambda) = (4\lambda + 2)^{|\mu|}$ and $w(x) \equiv 1$ also satisfies (3.16). Let us present direct and converse results. Set $\Delta_n(x) = \sqrt{1 - x^2}/n + 1/n^2$.

Theorem 3.3.4 (Ivanov, [163] and [166]). *Assume that w satisfies (3.16) with $C(\lambda) = \mathcal{O}(\lambda^c)$ $(\lambda \to \infty)$ for some $c > 0$.*

(i) *For every $k \geq 0$ and for every f with $f^{(k)} \in L_p[-1,1]$ we have*

$$E_{n+k}(f,w)_p \leq C(k)E_n(f^{(k)}, w(\Delta_n)^k)_p$$

and

$$E_{n+k}(f,w)_p \leq C(k)\,\tau_1(f^{(k)}, w(\Delta_n)^k, \Delta_n)_p$$

where

$$E_n(f,w)_p = \inf_{p \in \mathbb{P}_n} \|(f-p)w\|_{1,p}.$$

In particular,

$$E_{n+k}(f)_p \leq C(k)E_n(f^{(k)}, (\Delta_n)^k)_p$$

and

$$E_{n+k}(f)_p \leq C(k)\,\tau_1(f^{(k)}, (\Delta_n)^k, \Delta_n)_p.$$

(ii) *If for each $Q \in \Pi_m$, $m \leq n$,*

$$\|wQ^{(k)}(n\Delta_n)^k\|_p \leq C(k)\,m^k\|wQ\|_p, \tag{3.17}$$

then for every $r \in [1,p]$ and $f \in L_p[-1,1]$,

$$\tau_k(f, w, \Delta_n)_{r,p} \leq \frac{C(k)}{n^k}\sum_{j=0}^{n}(j+1)^{k-1}E_s(f,w)_p.$$

Using Koniagin [204] results we obtain

Corollary 3.3.5. *If $f \in L_p[-1,1]$ and $r \in [1,p]$ and $m \in \mathbb{N}_0$, then*

$$\tau_k(f, (n\Delta_n)^m, \Delta_n)_{r,p} \leq \frac{C(k,m)}{n^k}\sum_{j=0}^{n}(j+1)^{k-1}E_s(f,(n\Delta_n)^m)_p.$$

In particular

$$\tau_k(f, \Delta_n)_{r,p} \leq \frac{C(k)}{n^k}\sum_{j=0}^{n}(j+1)^{k-1}E_s(f)_p.$$

Corollary 3.3.6. *If $f \in L_p[-1,1]$, $r \in [1,p]$ and $0 < \alpha < 1$, one has*

$$E_n(f)_p = \mathcal{O}(n^{-\alpha}) \quad \Longleftrightarrow \quad \tau_k(f,\Delta(d))_{r,p} = \mathcal{O}(d^\alpha).$$

The direct estimate in Theorem 3.3.4 is given in terms of the first τ modulus of the derivative $f^{(k)}$. In [169] Ivanov presented the estimate in terms of the τ modulus of order k.

Fix $s > 0$ and let us denote for $W(s)$ the class of all weights $w \in C[-1, 1]$ that have the following properties: for each $x, t \in [-1, 1]$ with $\mid x - t \mid \leq \lambda \Delta_n(x)$,

$$0 < w(x) \leq C(\lambda)\, w(t)$$

and, in the case $\lambda \geq 1$,

$$0 < w(x) \leq C\lambda^s\, , w(t).$$

Theorem 3.3.7. *Suppose that $s > 0$ and $w \in W(s)$. If $k \in \mathbb{N}$ and $f \in L_p[-1, 1]$ $(1 \leq p \leq \infty)$, then*

$$E_{n+k}(f, w)_p \leq C(s, k)\, \tau_k(f, w, \Delta_n)_p.$$

Theorem 3.3.8. *Suppose that $s > 0$, $w \in W(s)$ and (3.17) holds. For $f \in L_p[-1, 1]$ $(1 \leq p \leq \infty)$, $0 < \alpha < k$, the following assertions are equivalent:*

(i) $E_n(f, w)_p = \mathcal{O}(n^{-\alpha})$,

(ii) $\tau_k(f, w, \Delta_n)_{1, p} = \mathcal{O}(n^{-\alpha})$.

In [173] Ivanov defined the τ modulus in a slightly different form:

$$\tau_k(f, w, \psi(t))_{r, p, [a, b]} \;=\; \| w(\cdot) \omega_k(f, \cdot, \psi(t, \cdot))_r \|_{p, [a, b]} \qquad (3.18)$$

where

$$\omega_k(f, x, \psi(t, x))_r = \left(\frac{1}{2\psi(t, x)} \int_{-\psi(t, x)}^{\psi(t, x)} \left| \Delta_v^k f(x) \right|^r dv \right)^{1/r}$$

and

$$\omega_k(f, x, \psi(t, x))_\infty = \sup \left\{ \left| \Delta_h^k f(x) \right| : \mid h \mid \leq \psi(t, x) \right\}.$$

Let us consider some types of weights. Two functions, v (continuous, strictly monotone, and $v(0) = 0$) and u, are associated with the weight w in neighborhoods of the end-points a and b. Let a and b be finite. Consider a neighborhood $[a, d]$ of a or $[d, b]$ of b; we write $v(x) = x/w(a + x)$ for $x \in (0, d - a]$ or $v(x) = x/w(b - x)$ for $\in (0, b - d]$, respectively. u is the inverse function to v, i.e., $u(v(x)) = v(u(x)) = x$. For $a = 0$ the functions u and w are connected by

$$u(x) = v(u(x)w(v(x)) = xw(u(x)).$$

Now consider the following classes:

Type 1. w is non-decreasing, v is strictly increasing in $[0, d]$, and $v(0) = \lim_{x \to 0} v(x) = 0$. For $0 < t \leq v(d)$ we set

$$\psi(t, x) = tw(x + u(t)). \qquad (3.19)$$

Type 2. w is non-increasing and unbounded in $(0, d]$ and, for every $x \in (0, d/2]$, satisfies the inequality

$$w(x) \leq A_2 w(2x).$$

In this case ψ is also defined by (3.19).

Type 3. v is non-increasing in $(0, d]$ and, for every $x \in (0, d]$, w satisfies the inequality $(t_0 = v(d)/2)$

$$w(x) \leq A_w(x - t_0 w(x)).$$

In this case we define

$$\psi(t, x) = tw(x).$$

The weight w will satisfy the following global condition:

There exist $A_5 \geq 1$, $a < d_3 < d_1 < d_2 < d_4 < b$, and weights w_1 in $[a, d_1]$ and w_2 in $[d_2, b]$ of some of the types described above such that $1/A_5 \leq w(x)/w_1(x) \leq A_5$ for $x \in [a, d_1]$, (3.20)
$1/A_5 \leq w(x)/w_2(x) \leq A_5$ for $x \in [d_2, b]$, and $1/A_5 \leq w(x) \leq A_5$ for $x \in [d_3, d_4]$.

Sometimes we shall require v to satisfy the additional conditions

$$\int_0^x v^k(y) \frac{dy}{y} \leq A_1 v^k(x), \qquad x \in (0, d] \tag{3.21}$$

and sometimes we shall require w to satisfy the additional conditions

$$u(\lambda x) \leq C(\lambda) u(x), \qquad \text{for any } x > 0, \ \lambda \geq 1, \quad \lambda x \leq d. \tag{3.22}$$

With v_j, u_j and ψ_j we denote the functions associated with the weight w_j, $j = 1, 2$. Then we set

$$\underline{\psi}(x, t) = \begin{cases} (2k)^{-1}\psi_1(t, x), & x \in [a, d_1], \\ (2k)^{-1}\psi_2(t, x), & x \in [d_2, b], \\ \text{linear and continuous}, & x \in [d_1, d_2]. \end{cases} \tag{3.23}$$

We also need the following condition:

There is $A > 1$ such that $1/A \leq \underline{\psi}(t, x)/\psi(t, x) \leq A$ for every $x \in [a, b]$ and the weights w_1 or w_2 from (3.20) satisfy (3.21) (3.24)
and (3.22) provided they are of Type 1.

Theorem 3.3.9 (Ivanov, [173]). *Let w satisfy (3.20) in $[a, b]$ and let ψ satisfy (3.24) for $0 < t \leq C(w)$. Then for every $f \in L_p[a, b] + W_p^k(w)$ we have*

$$C_1(k, w)\tau_k(f, \psi(t))_{p,p} \leq K(f, t^k, L_p, W_p^k(w)) \leq C_1(k, w)\tau_k(f, \psi(t))_{1,p}.$$

Let $w(x) = \sqrt{x(1-x)}$, $x \in [0, 1]$. We can choose $d_1 = 1/3$, $d_2 = 2/3$, $d_3 = 1/4$ and $d_4 = 3/4$, $w_1(x) = \sqrt{x}$, and $w_2(x) = \sqrt{1-x}$. Then $u_1(t) = t^2$ and $\psi_1(t, x) = t\sqrt{x + t^2}$. Therefore we can choose $\psi(t, x) = tw(x) + t^2$. Thus for $\varphi(x) = \sqrt{x(1-x)}$ the last theorem yields

$$C_1(k, w)\tau_k(f, \psi(t))_{p,p} \leq K(f, t^k, L_p, W_p^k(\varphi)) \leq C_1(k, w)\tau_k(f, \psi(t))_{1,p}.$$

Let w be symmetry in $[0, 1]$ (i.e., $w(1 - x) = w(x)$) and let w_1 from (3.20) be of Type 1 in $[0, 1/3]$ satisfying (3.21). Let us denote by u the function u_1, corresponding to w_1. We set

$$K^*(f, t^k, L_p, W_p^k(w)) = \inf \left\{ \|f - g\|_p + t^k \|w^k g^{(k)}\|_p + u(t)^k \|g^{(k)}\|_p \right\}.$$

Theorem 3.3.10 (Ivanov, [173]). *Under the above assumption we have*

$$K(f, t^k, L_p, W_p^k(w)) \sim K^*(f, t^k, L_p[0, 1], W_p^k(w)).$$

Results related with characterizations of the best approximation of the best approximation by algebraic polynomials in terms of the τ-modulus were given in [163], [164], [165], [167], [168], [169], [171], [170], and [172]. The extension to $L_p[-1, 1]$ with $0 < p < 1$ was given by Tachev in [367] and [368].

3.4 Ditzian-Totik moduli

In 1987 Ditzian and Totik published the book [102] where the following modulus was studied in detail.

Let $\Delta_h^r f(x)$ be the symmetric difference of order r (the difference is zero if some of the points are outside of the interval). For $1 \le p \le \infty$ and $f \in L_p[-1, 1]$ define

$$\omega_\varphi^r(f, t)_p = \sup_{0 < h \le t} \|\Delta_{h\varphi} f\|_p,$$

where $\varphi(x) = \sqrt{1 - x^2}$. Other functions φ can also be considered and it varies with the interval. For instance, we take $\varphi(x) = \sqrt{(x - a)(b - x)}$ for the interval $[a, b]$.

We remark that the ideas related with these moduli were developed by both authors in some previous papers (see [93], [95], [384], [385], [386] and other papers related with positive linear operators).

For continuous functions ($p = \infty$) it can be proved that the conditions

$$\sup_{0 < h \le t} \varphi^\alpha(x) \mid \Delta_h^r f(x) \mid = \mathcal{O}(t^\alpha)$$

and $\omega_\varphi^r(f, t)_\infty = \mathcal{O}(t^\alpha)$ $(t > 0)$ are equivalent ([92], [94], [384], [386]), but for $1 \le p < \infty$ these conditions are not equivalent [383].

In [173] Ivanov proved that the modulus (3.18) and the Ditzian-Totik ones are equivalent, for $1 \le p \le \infty$. Tachev verified the equivalence for $0 < p < 1$ [369]. Another proof was given by Ditzian, Hristov and Ivanov in [98].

From the point of view of applications in approximation theory it is a very important result connecting the weighted moduli with some K-functionals.

For $1 \le p \le \infty$ and $r \in \mathbb{N}$ define

$$K_{r,\varphi}(f, t^r)_p = \inf_g \left\{ \|f - g\|_p + t^r \|\varphi^r g^{(r)}\|_p : g^{(r-1)} \in A.C._{\cdot \text{loc}} \right\}$$

and

$$K_{r,\varphi}^*(f, t^r)_p = \inf_g \left\{ \|f - g\|_p + t^r \|\varphi^r g^{(r)}\|_p + t^{2r} \|g^{(r)}\|_p : g^{(r-1)} \in A.C._{\cdot \text{loc}} \right\}.$$

Here we present some results for $\varphi(x) = \sqrt{1-x^2}$, but see [102] for some other weight functions.

Theorem 3.4.1 (Ditzian and Totik, [102]). *For $1 \leq p \leq \infty$ and $r \in \mathbb{N}$, there exists a positive constant C_1, C_2 and t_0 such that, for all $f \in L_p[-1,1]$,*

$$C_1 \omega_\varphi^r(f,t) \leq K_{r,\varphi}(f,t^r)_p \leq K_{r,\varphi}^*(f,t^r)_p \leq C_2 \omega_\varphi^r(f,t), \qquad 0 < t \leq t_0.$$

3.4.1   Direct and converse results

Ditzian and Totik presented direct and converse results in terms of the modulus $\omega_\varphi^r(f,t)$.

Proposition 3.4.2. *Fix $1 \leq p \leq \infty$ and let λ be a positive integer. There exists a positive constant C with the following property. For $g \in A.C.[-1,1]$ such that $g' \in L_p[-1,1]$, $n \in \mathbb{N}$, $s = 2\lambda + 3$ and $m = 1 + [n/s]$, define*

$$L_{n,\lambda}(g,x) = \int_{-\pi}^{\pi} g(\cos(\arccos(x-t))) K_{m,s}(t)dt$$

where $K_{m,s}$ is given by (2.8). Then $L_{n,\lambda}(g) \in \mathbb{P}_n$ and

$$\|(\Delta_n)^\lambda(g - L_{n,\lambda}(g))\|_p \leq C \|(\Delta_n)^{\lambda+1} g'\|_p.$$

A proof of the last proposition can be found in [102] p. 80–82.

Theorem 3.4.3. *Let $\varphi(x) = \sqrt{1-x^2}$. For $1 \leq p \leq \infty$ and each $r \in \mathbb{N}$ there exist positive constants C_1 and C_2 such that, for all $f \in L_p[-1,1]$ and $n > r$,*

$$E_n(f)_p \leq C_1 \, \omega_\varphi^r(f,1/n)_p$$

and for $0 < t < 1$,

$$\omega_\varphi^r(f,t)_p \leq C_2 \, t^r \sum_{0 \leq n \leq 1/t} (n+1)^{r-1} E_n(f)_p.$$

Proof. Fix f. Taking into account Theorem 3.4.1, for each n we can find a function g_n such that

$$\|f - g_n\|_p + n^{-r}\|\varphi^r g_n^{(r)}\|_p + n^{-2r}\|g_n^{(r)}\|_p \leq 2K_{r,\varphi}^*(f,t^r)_p \leq C\omega_\varphi^r(f,1/n)_p.$$

Thus, it is sufficient to find a good approximant for g_n.

First, from Proposition 3.4.2 with $\lambda = r - 1$ and $g = g^{(r-1)}$, we obtain a polynomial $P_{n,1}$ such that

$$\|(\Delta_n)^{r-1}(g^{(r-1)} - P_{n,1})\|_p \leq C \|(\Delta_n)^r g^{(r)}\|_p \leq C_1 \left(\frac{1}{n^r}\|\varphi^r g^{(r)}\|_p + \frac{1}{n^{2r}}\|g^{(r)}\|_p \right)$$

$$\leq C_2 K_{r,\varphi}^* \left(f, \frac{1}{n^r} \right)_p.$$

Now we apply the same proposition (with $\lambda = r - 2$) to the function $g(u) = \int_0^u ((P_{n,1}(t) - g^{(r-1)}(t)dt$ to obtain a polynomial $P_{n,2} \in \mathbb{P}_{n+1}$ for which

$$\|(\Delta_n)^{r-2}(g^{(r-2)} - P_{n,2})\|_p \leq C \, \|(\Delta_n)^{r-1}[g^{(r-1)} - P_{n,1}]\|_p$$

$$\leq C_3 K^*_{r,\varphi}\left(f, \frac{1}{n^r}\right)_p.$$

Therefore, we can find a polynomial $P_{n,r} \in \mathbb{P}_{n+r-1}$ such that

$$\|g_n - P_{n,r}\|_p \leq C_r K^*_{r,\varphi}\left(f, \frac{1}{n^r}\right)_p.$$

Since

$$\|f - P_{n,r}\|_p \leq \|f - g_n\|_p + \|g_n - P_{n,r}\|_p,$$

we have the direct result.

For the converse results we use the Bernstein arguments, but now we use the Potapov inequality in Theorem 2.7.4. For $t \in (0, |)$, let $l = \max\{k \, 2^k \leq t\}$ and $\{P_n\}$ be the sequence of polynomials of the best approximation to f. From Theorems 3.4.1 and 2.7.4 one has

$$\omega^r_\varphi(f, t) \leq C K_{r,\varphi}(f, t^r)_p \leq C \left(\|f - P_{2^l}\|_p + t^r \|\varphi^r P^{(r)}_{2^l}\|_p \right)$$

$$= C \left(\|f - P_{2^l}\|_p + t^r \| \sum_{k=0}^{l-1} \|\varphi^r (P_{2^{k+1}} - P_{2^k})^{(r)}\|_p \right)$$

$$\leq C_1 \left(E_{2^l}(f)_p + t^r \sum_{k=0}^{l-1} 2^{(k+1)r} E_{2^k}(f)_p \right)$$

$$\leq C_2 t^r \sum_{0 \leq n \leq 1/t} (n+1)^{r-1} E_n(f)_p. \qquad \square$$

In particular, from the last result we obtain the following characterization.

Corollary 3.4.4. *For $0 < \alpha < r$ and $f \in L_p[-1, 1]$ the following assertions are equivalent:*

(i) $E_n(f)_p \leq C n^{-\alpha}$.

(ii) $\omega^r_\varphi(f, t)_p \leq C t^\alpha$.

The book contains different assertions concerning algebraic polynomials. The next result can be seen as an extension of an inequality due to Nikolskii and Stechkin.

Theorem 3.4.5. *Fix $f \in L_p[-1, 1]$ and $r \in \mathbb{N}$. Let P_n the best nth degree polynomial approximation to f in $L_p[-1, 1]$, then*

$$\|\varphi^r P^{(r)}_n\|_p \leq M \, \omega^r_\varphi(f, 1/n)_p,$$

where $\varphi(x) = \sqrt{1 - x^2}$ and M is independent of f and n.

For the converse they proved the following theorem.

Theorem 3.4.6. *Suppose that* $\|\varphi^r P_n^{(r)}\|_p \leq M\, n^r \psi(1/n)$, *where* P_n *is as in the last theorem and* $\psi(t) \to 0$ *and* $t \to 0$. *Then*

$$E_n(f) \leq M \int_0^{1/n} \frac{\phi(t)}{t}\,dt \qquad and \qquad \omega_\varphi^r(f,t)_p \leq M \int_0^t \frac{\phi(t)}{t}\,dt.$$

Corollary 3.4.7. *For* $0 < \alpha \leq r$ *and* $f \in L_p[-1,1]$ *the following assertions are equivalent:*

(i) $\|\varphi^r P_n^{(r)}\|_p \leq C n^{r-\alpha}$.
(ii) $\omega_\varphi^r(f,1/n)_p \leq C n^{-\alpha}$.

DeVore, Leviatan and Yu [90] extended the direct results in terms of the Ditzian-Totik modulus to L_p spaces, with $0 < p < 1$. Ditzian, Jiang and Leviatan proved the converse results [101]. For these last spaces the methods based on a K-functional do not work, as was shown by Ditzian, Hristov and Ivanov [98].

In 2008, Dai, Ditzian and Tikhonov extended (1.10) to the case of algebraic approximation.

Theorem 3.4.8 (Dai, Ditzian and Tikhonov, [80]). *For* $1 < p < \infty$, $s = \max\{p,2\}$ *and* $f \in L_p[-1,1]$, *one has*

$$t^r \left(\sum_{r \leq k \leq 1/t} k^{sr-1} E_k(f)_p^s \right)^{1/s} \leq C(r)\omega_\varphi^r(f,t)_p$$

and

$$t^r \left(\int_t^{1/2} \frac{\omega_\varphi^{r+1}(f,u)_p^s}{u^{rs+1}}\,du \right)^{1/s} + t^r E_r(f)_p \leq C(r)\omega_\varphi^r(f,t)_p.$$

For the best approximation in $C[-1,1]$, the Timan-type results are pointwise and Ditzian-Totik are in norm. Ditzian and Jiang presented a possible way to unify both theories.

Theorem 3.4.9 (Ditzian and Jiang, [99]). *For* $\lambda \in [0,1]$, $\varphi(x) = \sqrt{1-x^2}$, *there exists a constant* $C(r,\lambda)$ *such that, for all* $f \in C[-1,1]$ *there exists a sequence* $\{P_n\}$ *of polynomials such that,*

$$| f(x) - P_n(f,x) | \leq C(r,\lambda)\,\omega_{\varphi^\lambda}^r \left(f, \frac{1}{n} \left(\varphi(x) + \frac{1}{n} \right)^{1-\lambda} \right). \qquad (3.25)$$

If $\lambda = 0$, then we obtain the estimate in terms of the usual modulus of continuity and when $\lambda = 1$, we get the Ditzian-Totik estimate in norm. For the converse result Ditzian and Jiang proved the following. A similar result is not true for L_p spaces $(1 \leq p < \infty)$ and $0 \leq \lambda < 1$ (see [89] and [255]).

Theorem 3.4.10 (Ditzian-Jiang, [99]). *Fix $s > 0$ and let w be an increasing function satisfying*

$$w(\mu t) \leq C(\mu^s + 1)w(t). \tag{3.26}$$

If $f \in C[-1, 1]$ and there exists a sequence $\{P_n\}$ of polynomials such that

$$\mid f(x) - P_n(x) \mid \leq Cw\left(n^{-1}w(\delta_n^{1-\lambda}(x))\right), \tag{3.27}$$

then

$$\omega_{\varphi^\lambda}^r(f, t) \leq M\, t^r \sum_{0 < n \leq 1/t} n^{r-1}w(n^{-1}).$$

In order to obtain the converse results they need inequalities for the derivative of the polynomials in terms of the parameter λ.

Theorem 3.4.11 ([99]). *Suppose that for $P_n \in \mathbb{P}_n$ one has*

$$\mid P_n(x) \mid \leq M(n^{-1}\delta_n(x))^\beta w(n^{-1}\delta_n(x)^{1-\lambda}), \qquad \mid x \mid < 1,$$

where β is a real number and w satisfies (3.26). Then for $l \geq \beta + s(1 - \lambda)$,

$$\mid P_n^{(l)}(x) \mid \leq M_1(n^{-1}\delta_n(x))^{\beta - l}w(n^{-1}\delta_n(x)^{1-\lambda}), \qquad \mid x \mid < 1,$$

where M_1 depends on M, l, s, β and λ, but not on x, P_n or n.

3.4.2 Approximation in weighted spaces

Ditzian and Totik also considered approximations in weighted spaces. For a weight w the best approximation is defined by

$$E_n(f)_{p,w} = \inf_{P \in \mathbb{P}_n} \|w[f - P]\|_p.$$

The results are valid for some general weights, but the more important ones are the Jacobi weights $w(x) = (1 + x)^\alpha(1 - x)^\beta$.

The general class of weights J_p^* is defined as follows. $w \in J_p^*$ if

(a) $W(x) = w_-(\sqrt{1 + x})w_+(\sqrt{1 - x})$,
(b) $w_+(y) = y^{\gamma_1}v_+(y)$, $w_-(y) = y^{\gamma_1}v_-(y)$, where $\gamma_i > -2/p$ and $v_\pm(y) \sim 1$ on every interval $[\delta, \sqrt{2}]$, $\delta > 0$,
(c) for every $\varepsilon > 0$, $y^\varepsilon v_\pm(y)$ are increasing and $y^{-\varepsilon}v_\pm(y)$ are decreasing on $(0, \delta(\varepsilon))$ for some $\delta(\varepsilon) > 0$, and
(d) for $p = \infty$ we may have $\gamma_1 = 0$ or $\gamma_2 = 0$ in which case $v_-(y)$ or $v_+(y)$ have to be non-decreasing for small y.

For $f \in L_p[-1, 1]$ the main-part modulus is defined by

$$\Omega_\varphi^r(f, t)_{w,p} = \sup_{0 < h \leq t} \|w\Delta_{h\varphi}^r f\|_{p,[-1+2r^2h^2, 1-2r^2h^2]}.$$

Theorem 3.4.12. *For $w \in J_p^*$ and $\varphi(x) = \sqrt{1-x^2}$, we have*

$$E_n(f)_{w,p} \leq M \sum_{k=0}^{\infty} \Omega_\varphi^r(f, n^{-1}2^{-k})_{w,p}$$

and

$$\Omega_\varphi^r(f,h)_{w,p} \leq Mh^r \sum_{0 \leq n < 1/h} (n+1)^{r-1} E_n(f)_{w,p}.$$

For Jacobi weights we have a more general result.

Theorem 3.4.13. *If w is a Jacobi weight, then*

$$\omega_\varphi^r(f,h)_{w,p} \leq Mh^r \sum_{0 \leq n < 1/h} (n+1)^{r-1} E_n(f)_{w,p}.$$

The asymptotics of derivatives was also considered in the weighted case.

Theorem 3.4.14. *For $w \in J_p^*$ and P_n satisfying $\|w[f - P_n]\|_p = E_n(f)_{w,p}$ we have*

$$\|w\varphi^r P_n^{(r)}\|_p \leq Mn^r \int_0^{1/n} \frac{\Omega_\varphi^r(f,t)_{w,p}}{t} dt,$$

$$\Omega_\varphi^r(f,t)_{w,p} \leq M \sum_{k=1}^{\infty} 2^{-kr} n^{-r} \|w\varphi^r P_{2^k n}^{(r)}\|_p,$$

for $n = [1/t]$,

$$\|w\varphi^r P_n^{(r)}\|_p \leq M \sum_{k=0}^{n} (k+1)^{r-1} E_k(f)_{w,p}$$

and

$$E_n(f)_{w,p} \leq M \sum_{k=1}^{\infty} 2^{-kr} n^{-r} \|w\varphi^r P_{2^k n}^{(r)}\|_p.$$

In [80] one can find also results related with sharp inequalities in weighted space with Jacobi weights.

3.4.3 Marchaud inequalities

As we remarked above, Ditzian extended Marchaud inequality in [96], but the ideas he used were appropriated for studying weighted moduli of smoothness. The extension to algebraic approximation with the weight $\varphi(x) = \sqrt{1-x^2}$ was given by Totik.

Theorem 3.4.15 (Totik, [387]). *For $1 < p < \infty$, $q = \min\{p,2\}$ and $f \in L_p[-1,1]$, one has*

$$\omega_\varphi^r(f,1/n)_p \leq C(r,p)n^{-r} \left(\sum_{k=1}^{r} k^{rq-1} E_k(f)_p^q \right)^{1/q}.$$

Moreover, if $1 < p \leq 2$, then

$$\omega_\varphi^r(f,t)_p \leq C(r,p)\, t^r \left(\|f\|_p^p + \int_t^{1/2} \frac{\omega_\varphi^{r+1}(f,u)_p^p}{u^{rp+1}}\, du \right)^{1/p}.$$

This result also holds for other weight functions φ.

3.4.4 Simultaneous approximation

Ditzian and Jiang presented some results related with simultaneous approxima-
tion. In this section we use the notation $\varphi(x) = \sqrt{1-x^2}$ and $\delta_n(x) = n^{-1} + \varphi(x)$.

Theorem 3.4.16 (Ditzian-Jiang, [99]). *Fix $\lambda \in [0,1]$ and $f \in C[-1,1]$ and suppose there exists a sequence $\{P_n\}$ $(P_n \in \mathbb{P}_n)$ satisfying (3.27), where (3.26) holds
for w with $s = r$ and $\sum_{k=1}^{\infty} k^{r-1} w(k^{-1}) < \infty$. Then f has locally r continuous
derivatives and*

$$| \varphi^{r\lambda}(x)[f^{(r)}(x) - P_n^{(r)}(x)] | \leq M_1 \sum_{k > n\delta_n^{\lambda-1}} k^{r-1} w(k^{-1}).$$

Proof. It is known that, if ω is an increasing function, there and $\{u_k\}$ is an increasing sequence of positive numbers such that $2 \leq u_k/u_{k-1} \leq 4$, then there exists a
constant M such that

$$\sum_{k=1}^{l} u_k^r \omega(u_k^{-1}) \leq M \sum_{[u_l/2] \leq n \leq u_l} (n+1)^{r-1} \omega(n^{-1}).$$

Thus, if we set

$$u_i^{-1} = \frac{1}{2^i n} \left(\frac{1}{2^i n} + \varphi(x) \right)^{1-\lambda}$$

and consider the condition $\sum k^{r-1} \omega(k^{-1}) < \infty$, we prove that the series

$$f(x) = P_n(x) + \sum_{i=1}^{\infty} (P_{2^i n}(x) - P_{2^{i-1} n}(x))$$

converges. The equality holds because $P_n \to f$.

Taking into account Theorem 3.4.11, we know that there exists a constant
C_1 such that

$$\left| P_{2^i n}^{(r)}(x) - P_{2^{i-1} n}^{(r)}(x) \right| \leq C_1 \left(\frac{1}{2^i n} \delta_{2^i n}(x) \right)^{-r} \omega \left(\frac{1}{2^{i-1} n} (\delta_{2^{i-1} n}(x))^{1-\lambda} \right).$$

Therefore, the series

$$\sum_{i=1}^{\infty} \left(P_{2^i n}^{(r)}(x) - P_{2^{i-1} n}^{(r)}(x) \right)$$

converges uniformly locally in $(-1,1)$ and $(f - P_n)^{(r)}$ exists locally.

Finally, the estimate follows from the inequalities

$$\left| \varphi^{\lambda r}(x)(f - P_n)^{(r)}(x) \right| \leq \varphi^{\lambda r}(x) \sum_{i=1}^{\infty} \left| P_{2^i n}^{(r)}(x) - P_{2^{i-1} n}^{(r)}(x) \right|$$

$$\leq \varphi^{\lambda r}(x) \sum_{i=1}^{\infty} \left(\frac{1}{2^i n} \delta_{2^i n}(x) \right)^{-r} \omega \left(\frac{1}{2^i n} (\delta_{2^i n}(x))^{1-\lambda} \right)$$

$$\leq \sum_{i=1}^{\infty} \left(\frac{1}{2^i n} (\delta_{2^i n}(x))^{1-\lambda} \right)^{-r} \omega \left(\frac{1}{2^i n} (\delta_{2^i n}(x))^{1-\lambda} \right)$$

$$\leq M \sum_{k > n(\delta_n(x))^{\lambda-1}} k^{r-1} \omega \left(\frac{1}{k} \right). \qquad \square$$

Ditzian and Jiang presented a theorem for simultaneous approximation in L_p spaces that was not included in [102].

Theorem 3.4.17 ([99]). *Suppose $1 \leq p \leq \infty$, $f \in L_p[-1,1]$ and let $\{P_n\}$ $(P_n \in \mathbb{P}_n)$ be a sequence of polynomials satisfying $\|f - P_n\|_p = E_n(f)_p$. If*

$$\sum_{n=1}^{\infty} (n+1)^{r-1} E_n(f)_p < \infty,$$

then $f^{(r)}$ exists locally in the L_p sense and

$$\|\varphi^r[f^{(r)} - P_n^{(r)}]\|_p \leq M \sum_{k \geq n} (k+1)^{r-1} E_k(f)_p.$$

Proof. Let P_k be a polynomial of the best approximation of f in L_p and consider the series $\sum_{i=1}^{\infty} (P_{2^i n} - P_{2^{i-1} n})$. As in the proof of the last theorem, we obtain that the derivatives of the series exist locally. We use the inequality in (iv) of Theorem 2.7.4 to obtain

$$\left\| \varphi^r \sum_{i=1}^{\infty} (P_{2^i n}^{(k)} - P_{2^{i-1} n}^{(k)}) \right\|_p \leq \sum_{i=1}^{\infty} \left\| \varphi^r (P_{2^i n}^{(k)} - P_{2^{i-1} n}^{(k)}) \right\|_p$$

$$\leq C_1 \sum_{i=1}^{\infty} (2^i n)^r E_{2^i n}(f)_p \leq C_2 \sum_{k \geq n} (k+1)^{r-1} E_k(f)_p. \qquad \square$$

Theorem 3.4.18 ([99]). *Fix $\lambda \in [0,1]$ and $f \in C[-1,1]$ and suppose there exists a sequence $\{P_n\}$ $(P_n \in \mathbb{P}_n)$ satisfying*

$$| f(x) - P_n(x) | \leq C \omega_{\varphi^\lambda}^r \left(f, n^{-1} \delta_n^{1-\lambda}(x) \right).$$

Then

$$| \varphi^{r\lambda}(x) P_n^{(r)}(x) | \leq M_1 n^r \delta_n^{(\lambda-1)r}(x) \omega_{\varphi^\lambda}^r \left(f, n^{-1} \delta_n^{1-\lambda}(x) \right).$$

The analog of Theorem 2.8.11 and Corollary 5.6.15 in terms of $\omega_{\varphi^\lambda}^r$ moduli was obtained by Z. Ditzian, D. Jiang and D. Leviatan [100].

Theorem 3.4.19 (Ditzian, Jiang and Leviatan [100]). *Fix integers k, m and r and a real number $\lambda \in [0, 1]$. There exists a constant C such that, for each $f \in C^m[-1, 1]$ there exists a sequence of polynomials $P_n \in \mathbb{P}_n$ $(n \geq m + 1)$ for which*

$$| f^{(j)}(x) - Q_n^{(j)}(x) | \leq C \left(n^{-1}\varphi(x) \right)^{m-j} \omega_{\varphi^\lambda}^r \left(f^{(m)}, n^{-1}(\delta_n(x)^{1-\lambda}) \right), \qquad 0 \leq j \leq m)$$

and

$$| P_n^{(m+k)}(x) | \leq C\, n^k (\delta_n(x))^{-k} \omega_{\varphi^\lambda}^r \left(f^{(m)}, n^{-1}(\delta_n(x))^{1-\lambda} \right), \qquad k \geq r,$$

where $x \in [-1, 1]$.

For $r = 1, 2$ there are better estimates than those in the last theorem. In particular, Ditzian, Jiang and Leviatan showed that, for $r = 2$, the quantity $n^{-1} + \varphi(x)$ in (3.25) can be replaced by $\varphi(x)$.

Theorem 3.4.20 ([100]). *Fix $r \in \mathbb{N}$ and $\lambda \in [0, 1]$. There exists a constant C such that, for each $f \in C^m[-1, 1]$ there exists a sequence of polynomials $P_n \in \mathbb{P}_n$ for which*

$$| f^{(k)}(x) - P_n^{(k)}(x) | \leq C \left(n^{-1}\varphi(x) \right)^{m-k} \omega_{\varphi^\lambda}^l \left(f^{(m)}, n^{-1} \left(\varphi(x)\right)^{1-\lambda} \right),$$

for $l = 1, 2$ and $0 \leq k \leq m$ and

$$| P_n^{(m+k)}(x) | \leq C\, n^k\, (\delta_n(x))^{-k} \omega_{\varphi^\lambda}^l \left(f^{(m)}, n^{-1}(\varphi(x))^{1-\lambda} \right), \qquad k \geq l.$$

Kopotun provided a new proof of Theorem 3.4.9 and showed that the constant can be taken independent of λ.

Theorem 3.4.21 (Kopotun, [205]). *For any integer $r \geq 3$, there exists a constant $C(r)$ such that, for all $f \in C[-1, 1]$, each $\lambda \in [0, 1]$ and every $n \geq r - 1$, one can find a polynomial $P_n \in \mathbb{P}_n$ such that*

$$| f(x) - p_n(f, x) | \leq C(r)\, \omega_{\varphi^\lambda}^r \left(f, \frac{1}{n} \left(\varphi(x) + \frac{1}{n} \right)^{1-\lambda} \right), \qquad x \in [-1, 1].$$

Moreover, if $f \in C^1[-1, 1]$ then

$$| f'(x) - p_n'(f, x) | \leq C(r)\, \omega_{\varphi^\lambda}^{r-1} \left(f', \frac{1}{n} \left(\varphi(x) + \frac{1}{n} \right)^{1-\lambda} \right), \qquad x \in [-1, 1]$$

and, if $f \in C^2[-1, 1]$, then

$$| f''(x) - p_n''(f, x) | \leq C(r)\, \omega_{\varphi^\lambda}^{r-2} \left(f'', \frac{1}{n} \left(\varphi(x) + \frac{1}{n} \right)^{1-\lambda} \right), \qquad x \in [-1, 1].$$

This result only asserts simultaneous approximation up to the second derivative. Moreover, the theorem of Ditzian, Jiang and Leviatan is better near the endpoints of $[-1, 1]$, while the Kopotun is better on the interval $[-1 + n^{-2}, 1 - n^{-2}]$ (for the first and second derivatives).

In [206] Kopotun provided a different proof for the results of Li (Theorem 5.6.7)) and Ditzian, Jiang and Leviatan. He improved the estimates using a polynomial of a linear operator $P_n(f, x) : C^r[-1, 1] \to \mathbb{P}_n$, with the remarkable property that $P_n(f, x)$ is constructed independently of λ. The first theorem presented below improves the estimates inside the interval $[-1, 1]$, i.e., for $x \in [-1 + n^{-2}, 1 - n^{-2}]$.

Theorem 3.4.22 (Kopotun, [206]). *Let* $m \in \mathbb{N}_0$ *and* $r \in \mathbb{N}$. *Then for any* $n \geq m + r - 1$ *there exists a linear operator* $P_n : C^m[-1, 1] \to \mathbb{P}_n$ *such that for every* $\lambda \in [0, 1]$, $x \in [-1, 1]$ *and* $f \in C^m[-1, 1]$,

$$| f^{(k)}(x) - P_n^{(k)}(f, x) | \leq C(r, m) \, (\Delta_n(x)))^{j-k} \omega_{\varphi^\lambda}^{m+r-j} \left(f^{(j)}, n^{-\lambda} \, (\Delta_n(x))^{1-\lambda} \right),$$

for $0 \leq k \leq m$ *and any* $j \in \mathbb{N}$ *satisfying* $k \leq j \leq m$. *Also, the following estimates hold for every* $\lambda \in [0, 1]$ *and* $x \in [-1, 1]$:

$$| P_n^{(k)}(f, x) | \leq C(k) \, (\Delta_n(x))^{j-k} \, \omega_{\varphi^\lambda}^{m+r-j} \left(f^{(j)}, n^{-\lambda} \, (\Delta_n(x))^{1-\lambda} \right),$$

for $k \geq m + r$ *and any* $j \in \mathbb{N}_0$, $0 \leq j \leq r$.

In particular, by taking $\lambda = 0$ and $j = k$ for the first inequality and $j = 0$ for the second inequality one has

Corollary 3.4.23 ([206]). *For* $f \in C^m[-1, 1]$, $r \in \mathbb{N}$ *and any* $n \geq m + r - 1$ *a linear operator* $P_n : C^m[-1, 1] \to \mathbb{P}_n$ *exists such that for* $x \in [-1, 1]$,

$$| f^{(k)}(x) - P_n^{(k)}(f, x) | \leq C(k) \omega_{m+r-k} \left(f^{(k)}, (\Delta_n(x)) \right),$$

for $0 \leq k \leq m$ *and*

$$| P_n^{(k)}(f, x) | \leq C(k) \, (\Delta_n(x))^{-k} \, \omega_{m+r} \left(f, \Delta_n(x) \right),$$

for $k \geq m + r$.

Kopotun also presented a complicated result which improves near the endpoints.

Theorem 3.4.24 ([206]). *Let* $m \in \mathbb{N}_0$, $r \in \mathbb{N}$ *and* $k_0 \geq m + r$. *Then for any* $n \geq \max\{m+r-1, 2m+1\}$ *there exists a linear operator* $P_n : C^m[-1, 1] \to \mathbb{P}_n$ *such that for every sequence* $\{\alpha_k\}_{k=0}^m \subset [1/r, 1]$, $\lambda \in [0, 1]$, $0 \leq k \leq r$ *and* $f \in C^m[-1, 1]$,

$$| f^{(k)}(x) - P_n^{(k)}(f, x) | \leq C(k_0) \, (\Delta_n(x)))^{m-k} \omega_{\varphi^\lambda}^m \left(f^{(m)}, n^{-\lambda} \, (\Delta_n(x))^{1-\lambda} \right),$$

for $x \in [-1, -1 + n^{-2}] \cup [1 - n^{-2}, 1]$, and

$$| f^{(k)}(x) - P_n^{(k)}(f, x) | \leq C(k_0) n^{2 - 2\alpha_k r} (1 - x^2)^{m-k+1-\alpha_k r}$$
$$\times \omega_{\varphi^\lambda}^r \left(f^{(j)}, n^{-\lambda}((1 - x^2)^{\alpha_k} n^{2 - 2\alpha_k r})^{1-\lambda} \right),$$

for $x \in [-1 + n^{-2}, 1 - n^{-2}]$.

Also, there exists a constant $n_0 = n_0(k_0)$ such that if $n \geq n_0$, then for every $\{\alpha_k\}_{k=m+r}^{k_0} \subset [1/r, 1]$, $\{r_k\}_{k=m+r}^{k_0} \subset [0, k_0]$, and for $\lambda \in [0, 1]$ and $m + r \leq k \leq k_0$, operator P_n satisfies

$$| P_n^{(k)}(f, x) | \leq C(k_0) (\Delta_n(x)))^{m-k} \omega_{\varphi^\lambda}^r \left(f^{(r)}, n^{-\lambda} (\Delta_n(x))^{1-\lambda} \right),$$

for $x \in [-1 + n^{-2}, 1 - n^{-2}]$, and

$$| P_n^{(k)}(f, x) | \leq C(k_0) n^{2(r_k - m + k + 1 - \alpha_k r)} (1 - x^2)^{r_k + 1 - \alpha_k r}$$
$$\times \omega_{\varphi^\lambda}^r \left(f^{(j)}, n^{-\lambda}((1 - x^2)^{\alpha_k} n^{2 - 2\alpha_k r})^{1-\lambda} \right),$$

for $x \in [-1, -1 + n^{-2}] \cup [1 - n^{-2}, 1]$.

Some important corollaries follow from the last theorem.

Corollary 3.4.25. *Fix $r \in \mathbb{N}$. Then for any $n \geq \max\{m + r - 1, 2m + 1\}$ there exists a linear operator $P_n : C^m[-1, 1] \to \mathbb{P}_n$ such that for every $0 \leq k \leq m$, the following inequalities hold:*

$$| f^{(m)}(x) - P_n^{(m)}(f, x) | \leq C(r, m) \Delta_n^{m-k} \omega_r(f^{(r)}, \Delta_n(x))$$

for $x \in [-1 + n^{-2}, 1 - n^{-2}]$, and

$$| f^{(k)}(x) - P_n^{(k)}(f, x) | \leq C(r, m) \Gamma_{nrmk}^{m-k}(x) \omega_r(f^{(m)}, \Gamma_{nrmk}(x)),$$

for $x \in [-1, -1 + n^{-2}] \cup [1 - n^{-2}, 1]$, where

$$\Gamma_{nrmk}(x) := (1 - x^2)^{(m-k+1)/(m-k+r)} (1/n^2)^{(r-1)/(m-k+r)}.$$

Moreover, these estimates are exact in the sense that for no $0 \leq k \leq m$ can $\Gamma_{nrmk}(x)$ be replaced by $(1 - x^2)^{\alpha_k} n^{2\alpha_k - 2}$ with $\alpha_k > (m - k + 1)/(m - k + r)$.

Notice that $\Gamma_{nrmk}(x) \leq \sqrt{1 - x^2}/n$ for any $0 \leq k \leq m + 2 - r$ and for all $x \in [-1, -1 + n^{-2}] \cup [1 - n^{-2}, 1]$. The inequalities in the last theorem hold for all $0 \leq k \leq m$, while Theorem 5.6.5 may not be true if $k > m + 2 - r$. It is also of interest to consider the special case $r = 1$ in the corollary.

Corollary 3.4.26. *For any $n \geq 2m + 1$ there exists a linear operator $P_n : C^m[-1, 1] \to \mathbb{P}_n$ such that for every $0 \leq k \leq m$, $x \in [-1, 1]$ a function $f \in C^m[-1, 1]$, the following inequality holds:*

$$| f^{(k)}(x) - P_n^{(k)}(f, x) | \leq C(m) \Gamma_n(x)^{m-k} \omega(f^{(m)}, \Gamma_n(x)),$$

where $\Gamma_n(x) = \min\{1 - x^2, \sqrt{1 - x^2}/n\}$. Moreover, $\Gamma_n(x)$ cannot be replaced by $\min\{(1 - x^2)^\alpha, \sqrt{1 - x^2}/n\}$ with $\alpha > 1$.

By using the arguments of Leviatan, Kopotun also obtained an estimate in terms of the best approximation which improved those given by Kilgore in (2.40).

Corollary 3.4.27. *Then for any $n \geq 2m+1$ and $f \in C^m[-1,1]$ there is a polynomial $P_n \in \mathbb{P}_n$ such that for every $0 \leq k \leq m$ and $x \in [-1,1]$,*

$$| f^{(k)}(x) - P_n^{(k)}(x) | \leq C(r) \left(\min\left\{ 1 - x^2, \frac{\sqrt{1-x^2}}{n} \right\} \right)^{m-k} E_{n-m}(f^{(m)}).$$

3.4.5 A Banach space approach

In [48] Butzer, Jansche and Stens considered the problem of generalizing the ideas of Butzer and Scherer ([51] and [52]) in such a way that we can also obtain the results of Ditzian and Totik. Solving the problem is justified by reasons of economy: to avoid many tricky and technical arguments.

The main idea was to use Jackson-type inequalities and K-functionals with respect to a family of seminorms instead of a single seminorm. In particular they proved the Lipschitz spaces associated with Ditzian-Totik moduli coincide with the ones obtained by means of the Jacobi transform.

For $\gamma > 0$, let $\Phi(\gamma)$ be the class of all functions $\phi : (0,1] \to \mathbb{R}$ such that $0 < \phi(s) \leq \phi(t) \leq \phi(1) < \infty$, for $0 < s < t \leq 1$, $\lim_{t \to 0+} \phi(t) = 0$ and

$$\int_t^1 \frac{\phi(u)}{u^{1+\gamma}} du = \mathcal{O}\left(\frac{\phi(t)}{t^\gamma} \right).$$

Let us present the general theorem. We use the following notation:

$$K(f, t, X, Y) = \inf_{g \in Y} \{ \|f - g\|_X + t \, | \, g \, |_Y \}$$

and

$$K^*(f, t, X, Y) = \sup_{0 < h \leq t} \inf_{g \in Y} \{ | \, f - g \, |_{X(h)} + t \, | \, g \, |_Y \}.$$

Theorem 3.4.28 (Butzer, Jansche and Stens, [48]). *Fix $\gamma > 0$, $M > 0$ and $n_0 \in \mathbb{N}$. Let X be a normed space with norm $\| \cdot \|_X$ and $Y \subset X$ a linear subspace with a seminorm $| \cdot |_Y$. Let $\{| \cdot |_{X(t)}\}_{t \in (0,1]}$ be a family of seminorms on X satisfying*

$$| f \, |_{X(t)} \leq | f \, |_{X(s)} \leq M \|f\|_X, \qquad (0 < s \leq t \leq 1), \tag{3.28}$$

for a constant M, independent of f, s and t, and if $\{f_n\}$ is a Cauchy sequence in X with $\lim_{n \to \infty} | f_n |_{X(t)} = 0$ for all $t \in (0,1]$, then

$$\lim_{n \to \infty} \|f_n\|_X = 0. \tag{3.29}$$

Let $\{M_n\}_{n=0}^{\infty}$ be a sequence of linear manifolds in X such that

$$M_n \subset M_{n+1} \subset Y, \qquad\qquad (n \in \mathbb{N}_0),$$
$$|\,g_n\,|_Y \leq M n^{\gamma} \|g_n\|_X, \qquad (g_n \in M_n, \; n \in \mathbb{N}_0),$$
$$\|g_n\|_X \leq M \,|\,g_n\,|_{X(2/n)}, \qquad (g_n \in M_n, \; n \geq n_0), \qquad (3.30)$$

and

$$E_n(f, X(1/n)) \leq M \, n^{-\gamma} \,|\,f\,|_Y, \qquad (f \in Y, \; n \geq n_0), \qquad (3.31)$$

where

$$E_n(f, X(1/n)) = \inf_{g \,\in\, M_n} \,|\,f - g\,|_{X(1/n)}\,.$$

(a) *For $\phi \in \Phi(\gamma)$ and $f \in X$ the following assertions are equivalent:*

   (i) $E_n(f) = \displaystyle\inf_{g \,\in\, M_n} \|f - g\|_X = \mathcal{O}(\phi(1/n)), \quad (n \to \infty).$

   (ii) $K(f, t^{\gamma}, X, Y) = \mathcal{O}(\phi(t)), \quad (t \to 0).$

(b) *If*

$$\int_0^t \frac{\phi(u)}{u} du = \mathcal{O}\left(\phi(t)\right), \qquad (3.32)$$

then (i) *and* (ii) *are further equivalent to*

(ii)* $K^*(f, t^{\gamma}, X, Y) = \mathcal{O}\left(\phi(t)\right), \quad (n \to \infty).$

(c) *Assume that* (3.32) *holds. If for each $f \in X$ and $n \in \mathbb{N}_0$, there exists $g_n(f) \in M_n$ such that $E_n(f) = \|f - g_n(f)\|$ and $\lim_{n\to\infty} E_n(f) = 0$, then the assertions given above are equivalent to*

   (iii) $|\,g_n(f)\,|_Y = \mathcal{O}(n^{\gamma} \phi(1/n)), \quad (n \to \infty).$

(d) *Fix $\delta > 0$ and assume that the conditions in* (a) *hold. Let $Z \subset X$ $(M_n \subset Z)$ be a subspace with a seminorm $|\cdot|_Z$ such that Z is a Banach space under the norm $\|\cdot\|_Z = \|\cdot\|_X + |\cdot|_Z,$*

$$E_n(f, X(1/n)) \leq M n^{-\delta} \,|\,f\,|_Z, \qquad (f \in Z, \; n \geq n_0),$$

and

$$|\,g_n\,|_Z \leq M \, n^{\delta} \|g_n\|_X, \qquad (g_n \in M_n, \; n \in \mathbb{N}_0).$$

If

$$\int_0^t \frac{\phi(u)}{u^{1+\delta}} du = \mathcal{O}\left(\frac{\phi(t)}{t^{\delta}}\right), \qquad (3.33)$$

then the assertion (i) *is equivalent to*

   (iv) $f \in Z, \quad |\,f - g_n(f)\,|_Z = \mathcal{O}(n^{\delta} \phi(1/n)), \quad (n \to \infty),$

     $f \in Z, \quad E_n(f, Z) = \displaystyle\inf_{g \,\in\, M_n} |\,f - g\,|_Z = \mathcal{O}(n^{\delta} \phi(1/n)), \quad (t \to 0).$

Finally if $\phi \in \Phi(\gamma)$ all the assertions given above are equivalent.

We only present a proof for (a), (b) and (c).

Proof. (a) ((i) $\implies$ (ii)). Fix $t \in (0,1]$ and $k \in \mathbb{N}_0$ such that $2^{-k-1} < t \le 2^{-k}$. In order to simplify, we assume that each M_n is an existence set. That is, for each $n \in \mathbb{N}$, there exists $P_n \in M_n$ such that $E_n(f) = \|f - P_n\|_X$.

From the Bernstein-type inequality ($|\,g_n\,|_Y \le Mn^\gamma\|g_n\|_X$), we know that

$$|\,P_{2^k}\,|_Y = \left|\,P_1 + \sum_{j=1}^{k}(P_{2^j} - P_{2^{j-1}})\,\right|_Y \le C_1\|P_1\|_X + M\sum_{j=1}^{k}2^{j\gamma}\|(P_{2^j} - P_{2^{j-1}})\|_X$$

$$\le C_2\left(\|f\|_X + \sum_{j=0}^{k}2^{j\gamma}E_{2^j}(f)\right) \le C_3\left(\|f\|_X + \sum_{j=0}^{k}2^{j\gamma}\phi(2^{-j})\right).$$

Taking into account that $M_n \subset Y$ and $t^\gamma \sum_{j=0}^{k} 2^{j\gamma}\phi(2^{-j}) \le C_4\phi(t)$ (for $\phi \in \Phi(\gamma)$), one has

$$K(f,t^\gamma,X,Y) \le \|f - P_{2^k}\|_X + t^\gamma\,|\,P_{2^k}\,|_Y$$

$$\le C_5\left(\phi(2^{-k}) + t^\gamma\|f\|_X + t^\gamma\sum_{j=0}^{k}2^{j\gamma}\phi(2^{-j})\right) \le C_6\phi(t).$$

((ii) $\implies$ (i)) Let us first verify a Jackson-type inequality. Fix $g \in Y$ and $\varepsilon > 0$. For $n \ge n_0$, take elements $Q_{n2^k} \in M_{2^k n}$ such that $|\,Q_{2^k n} - g\,|_{X(2^{-k}n^{-1})} \le E_{2^k n}(g, X(2^{-k}n^{-1})) + \varepsilon/2^k$. From (3.28) and (3.30), we know that

$$\|Q_{2^{k+1}n} - Q_{2^k n}\|_X \le M\left(|\,Q_{2^{k+1}n} - g\,|_{X(2^{-k-1}n^{-1})} + |\,g - Q_{2^k n}\,|_{X(2^{-k}n^{-1})}\right)$$

$$\le M\left(E_{2^{k+1}n}(g, X(2^{-k-1}n^{-1})) + E_{2^k n}(g, X(2^{-k}n^{-1})) + \varepsilon 2^{-k}\right).$$

Therefore

$$\sum_{k=0}^{\infty} \|Q_{2^{k+1}n} - Q_{2^k n}\|_X \le M\left(\sum_{k=0}^{\infty} E_{2^k n}(g, X(2^{-k}n^{-1})) + \varepsilon\right).$$

Thus $\{Q_{2^k n}\}$ is a Cauchy sequence in X and, for $t \in (0,1]$,

$$\lim_{k\to\infty} |\,g - Q_{2^k n}\,|_{X(t)} \le \lim_{k\to\infty} |\,g - Q_{2^k n}\,|_{X(2^{-k}n^{-1})}$$

$$\le \lim_{k\to\infty} M\left(E_{2^k n}(g, X(2^{-k}n^{-1})) + \varepsilon 2^{-k}\right) = 0.$$

From (3.29) one has $\lim_{k\to\infty} \|g - Q_{2^k n}\|_X = 0$ and there holds the representation

$$g - g_n = \sum_{k=0}^{\infty}(g_{2^{k+1}n} - g_{2^k n}),$$

where the convergence is considered with respect to the norm of X.

Now

$$E_n(g) \le \|g - g_n\|_X \le \sum_{k=0}^{\infty} \|g_{2^{k+1}n} - g_{2^k n}\|_X \le M\left(\varepsilon + \sum_{k=0}^{\infty} E_{2^k n}(g, X(2^{-k}n^{-1}))\right).$$

Since $\varepsilon > 0$ is arbitrary we obtain (see (3.31))

$$E_n(g) \le M \sum_{k=0}^{\infty} E_{2^k n}(g, X(2^{-k}n^{-1})) \le C_1 \sum_{k=0}^{\infty} 2^{-k\gamma} n^{-\gamma} \mid g \mid_Y \le C_2 n^{-\gamma} \mid g \mid_Y.$$

With this Jackson-type inequality (i) follows easily, since

$$E_n(f) \le E_n(f - g) + E_n(g) \le \|f - g\|_X + C_2 n^{-\gamma} \mid g \mid_Y$$

and $g \in Y$ is arbitrary.

(b) (ii) $\Longrightarrow$ (ii)*. It follows from the inequality

$$K^*(f, t^\gamma, X, Y) \le K(f, t^\gamma, X, Y).$$

(ii)* $\Longrightarrow$ (i). Fix any $g \in Y$. For $n \ge n_0$,

$$E_n(f, X(1/n)) \le E_n(f - g, X(1/n)) + E_n(g, X(1/n))$$
$$\le M\left(\mid f - g \mid_{X(n^{-1})} + n^{-\gamma} \mid g \mid_Y\right).$$

Since $g \in Y$ is arbitrary, one has

$$E_n(f) \le MK^*(f, n^{-\gamma}, X, Y) \le C\phi(n^{-1}).$$

(c) (i) $\Longrightarrow$ (iii). Fix $n \in \mathbb{N}$ and $k \in \mathbb{N}_0$ such that $2^k \le n < 2^{k+1}$. Set $g_n = g_n(f)$. Taking into account the Bernstein-type inequality, (i) and the properties of ϕ, we obtain

$$\mid g_n \mid_Y = \left| g_1 + \sum_{j=1}^{k} (g_{2^j} - g_{2^{j-1}}) + (g_n - g_{2^k}) \right|$$

$$\le C_1 \left(\|g_1\|_X + \sum_{j=1}^{k} 2^{j\gamma} \|g_{2^j} - g_{2^{j-1}}\|_X + n^\gamma \|g_n - g_{2^k}\|_X \right)$$

$$\le C_2 \left(\|f\|_X + \|f - g_1\|_X \right.$$
$$\left. + \sum_{j=0}^{k} 2^{j\gamma} \|f - g_{2^j}\|_X + n^\gamma \|g_n - f\|_X + 2^{(k+1)\gamma} \|f - g_{2^k}\|_X \right)$$

$$\le C_3 \left(\|f\|_X + n^\gamma E_n(f) + \sum_{j=0}^{k} 2^{j\gamma} E_{2^j}(f) \right).$$

$$\le C_4 \left(\|f\|_X + n^\gamma \phi(n^{-1}) + \sum_{j=0}^{k} 2^{j\gamma} \phi(2^{-j}) \right) \le C_3 n^\gamma \phi(n^{-1}).$$

(iii) $\Longrightarrow$ (i). From the Jackson-type inequality we know that, for $n \geq n_0$ and $k \in \mathbb{N}_0$

$$E_{2^k n}(f) \leq E_{2^k n}(f - g_{2^{k+1}n}) + E_{2^k n}(g_{2^{k+1}n})$$

$$\leq \|f - g_{2^{k+1}n}\|_X + M(2^k n)^{-\gamma} \mid g_{2^{k+1}n} \mid_Y = E_{2^{k+1}n}(f) + M(2^k n)^{-\gamma} \mid g_{2^{k+1}n} \mid_Y .$$

Since $\|f - g_n\|_X \to 0$, condition (iii) yields

$$E_n(f) = \sum_{k=0}^{\infty} (E_{2^k n}(f) - E_{2^{k+1}n}(f))$$

$$\leq M \sum_{k=0}^{\infty} (2^k n)^{-\gamma} (2^{k+1}n)^{\gamma} \phi(2^{-k-1}n^{-1}) \leq C\phi(n^{-1}). \qquad \square$$

Let us show how this result can be applied in weighted approximation. For $\alpha, \beta > -1$ and $1 \leq p < \infty$ let $L_p^{(\alpha,\beta)}[-1,1]$ be the space of all f such that

$$\|f\|_{p,(\alpha,\beta)} = \left(\int_{-1}^{1} \mid f(u) \mid^p w_{\alpha,\beta}(u) du \right)^{1/p} < \infty,$$

where

$$w_{\alpha,\beta}(x) = (1-x)^{\alpha}(1+x)^{\beta}.$$

For $\alpha, \beta \geq 0$, $C_{\alpha,\beta}[-1,1]$ is the space of all continuous functions, for which the limits $\lim_{x \to -1} w_{\alpha,\beta}(x)f(x)$ and $\lim_{x \to 1} w_{\alpha,\beta}(x)f(x)$ exist, with the norm

$$\|f\|_{\infty,(\alpha,\beta)} = \sup_{x \in [-1,1]} \mid w_{\alpha,\beta}(x)f(x) \mid .$$

We need the family of seminorms used in Theorem 3.4.28. In order to apply this theorem we fix $c > 0$ and, for $t \in (0, 1/\sqrt{c})$, set

$$\mid f \mid_{X(t,c)} = \left(\int_{I(t,c)} \mid f(u) \mid^p w_{\alpha,\beta}(u) du \right)^{1/p}, \qquad f \in L_p^{(\alpha,\beta)}[-1,1]$$

and

$$\mid f \mid_{X(t,c)} = \sup_{x \in I(t,c)} \mid w_{\alpha,\beta}(x)f(x) \mid, \qquad f \in C_{\alpha,\beta}[-1,1],$$

where $I(t,c) = [-1+ct^2, 1-ct^2]$. Moreover, for $1/\sqrt{c} \leq t \leq 1$, we set $\mid f \mid_{X(t,c)} = 0$.

In the following we omit the interval $[-1,1]$ in the notation and X will be any one of the spaces defined above.

Let us denote $\varphi(x) = \sqrt{1-x^2}$ and consider the differential operator

$$(D^s f)(x) = \varphi^s(x)f^{(s)}(x), \qquad (x \in (-1,1), \quad s \in \mathbb{N}_0). \qquad (3.34)$$

The associated Sobolev spaces are given by

$$W^s_{p,(\alpha,\beta)} = \left\{ f \in L_p^{(\alpha,\beta)}[-1,1] : f = F \text{ a.e., } F \in AC^s_{\text{loc}}, D^s F \in L_p^{(\alpha,\beta)} \right\},$$

and

$$W^s_{\infty,(\alpha,\beta)} = \left\{ f \in C^{s-1}[-1,1] : f = F \text{ a.e., } f \in C^{s-1}(-1,1), D^s f \in C_{\alpha,\beta} \right\}.$$

It can be proved that the linear operator $D^s : W^s_X \to X$ is closed. Thus W^s_X is a Banach space with respect to the norm

$$\|f\|_{W^s_X} = \|f\|_X + \|D^s f\|_X, \qquad (f \in W^s_X).$$

Also in W^s_X we consider the seminorm $| f |_{W^s_X} = \|D^s f\|_X$. In this case the Jackson-type inequality (3.31) can be proved in the form

$$E_n(g, X(1/n), c) \leq M\, n^{-s} \|D^s g\|, \qquad (g \in W^s_X),$$

(see Proposition 5.1 in [48], the proof follows some ideas of Ditzian and Totik [102]).

The needed Bernstein-type inequality had been proved in 1974 by Khalilova [188]. For $s \in \mathbb{N}$,

$$\|D^s p_n\|_X \leq M\, n^s \|p_n\|_X, \qquad (p_n \in \mathbb{P}_n, \ n \in \mathbb{N}_0),$$

where the constant M is independent of n.

Finally we need the inequality (3.30). But it can be proved that, for all $c > 0$ there exists a constant $M > 0$ such that, for all $n \in \mathbb{N}$, $n > \sqrt{2}c$,

$$\|p_n\|_X \leq M\, | p_n |_{X(1/n,c)}, \qquad (p_n \in \mathbb{P}_n).$$

A proof can be found in the Nevai book [268]. Another proof was given in the Ditzian and Totik book (Chapter 8.4 of [102]).

Now we have all the necessary ingredients in order to use Theorem 3.4.28. But we first present a notion introduced by Ditzian and Totik. For $s \in \mathbb{N}$ and $f \in X$ the weighted main-part modulus is defined by

$$\Omega_s(f, t, X) = \sup_{0 < h \leq t} | \overline{\Delta}^s_{h\varphi(x)} f(x) |_{X(h, 2s^2)}, \qquad (0 < t < 1/(\sqrt{2}s)),$$

where $\overline{\Delta}^s_h$ denotes the central difference of order s,

$$\overline{\Delta}^s_j f(x) \sum_{k=0}^{s} (-1)^k \binom{s}{k} f \left(x + \left(\frac{s}{2} - k \right) h \right).$$

Theorem 3.4.29. *Fix $s, r \in \mathbb{N}_0$ and a real σ with $r < \sigma < s$. Moreover set $\varphi(x) = \sqrt{1-x^2}$ and let the operator D^s be defined by (3.34). Fix $f \in X$ and let $\{p_n(f)\}$ be the sequence of polynomials of the best approximation to f in the norm of X.*

The following assertions are equivalent:

(i) $$E_n(f, X) = \mathcal{O}(n^{-\sigma}), \qquad\qquad (n \to \infty),$$

(ii) $$K(f, t^s, X, W_X^s) = \mathcal{O}(t^\sigma), \qquad\qquad (t \to 0),$$

(iii) $$K^*(f, t^s, X, W_X^s) = \mathcal{O}(t^\sigma), \qquad\qquad (t \to 0),$$

(iv) $$\Omega_s(f, t, X) = \mathcal{O}(t^\sigma), \qquad\qquad (t \to 0),$$

(v) $$\|D^s p_n(f)\|_X = \mathcal{O}(n^{s-\sigma}), \qquad\qquad (n \to \infty),$$

(vi) $$f \in W_X^r \quad \text{and} \quad \|D^r f - D^r p_n(f)\|_X = \mathcal{O}(n^{r-\sigma}), \qquad (n \to \infty),$$

(vii) $$f \in W_X^r \quad \text{and} \quad \inf_{p \in \mathbb{P}_n} \|\varphi^r[f^{(r)} - p]\|_X = \mathcal{O}(n^{r-\sigma}), \qquad (n \to \infty).$$

The equivalence of the first five assertions was first given by Ditzian and Totik [102] (Chapter 8). For $X = C[-1, 1]$ and the relations (i) $\Leftrightarrow$ (v) see also the papers of Golischek [143], Scherer and Wagner [330] and Stens ([348] and [349]). The equivalence (i) $\Leftrightarrow$ (v) $\alpha = \beta = 0$ is due to Heilmann [158]. The work [48] also includes some results related with characterizations when the function $\phi(t) = t^\sigma$ is replaced by $\psi(t) = t^\sigma(1 - \log t)$ or $\psi(t) = e^{-1/t}$. The passage from K-functional to moduli of smoothness is a complicated task.

For $f \in X$ the Jacobi transform is defined by

$$\widehat{f}(k) = \int_{-1}^{1} f(u) R_k^{(\alpha, \beta)}(u)\, w_{(\alpha, \beta)}(u)du, \qquad (k \in \mathbb{N}_0),$$

where

$$R_k^{(\alpha, \beta)}(x) = \frac{P_k^{(\alpha, \beta)}(x)}{P_k^{(\alpha, \beta)}(1)}$$

is the normalized Jacobi polynomial of degree k. The generalized translation operator is defined in terms of its Jacobi transform

$$[\tau_t f]^\wedge(k) = R_k(t)\widetilde{f}(k), \qquad (k \in \mathbb{N}_0, \quad t \in (-1, 1), \quad f \in X).$$

From Gasper [134], we know that the translation is a bounded linear operator mapping X into itself and satisfying

$$\|\tau_t f\|_X \leq M \|f\|, \qquad (t \in (-1, 1), \quad f \in X),$$

and

$$\lim_{t \to 0+} \|\tau_t f - f\|_X = 0, \qquad (f \in X),$$

if and only if $\alpha \geq \beta > -1$ and $\alpha + \beta \geq -1$. Now define

$$\Delta_t^J(f) = f - \tau_t f.$$

For $s \in \mathbb{N}$ the modulus of smoothness of $f \in X$ is defined by

$$\omega_s^J(f, t, X) = \sup\left\{ \|\Delta_{h_1}^J \Delta_{h_2}^J \cdots \Delta_{h_s}^J f\|_X : 1 - t \leq h_i \leq 1, \ i = 1, 2, \ldots, s \right\}.$$

Notice that in contrast to the classical moduli the increments h_i are allowed to be different in each iteration. It was proved in [48] that for $0 < \sigma < s$ and $f \in X$,

$$w_s^J(f, t, X) = \mathcal{O}(t^\sigma) \quad \Longleftrightarrow \quad \Omega_{2s}(f, t, X) = \mathcal{O}(t^{2\sigma}),$$

if $X = L_p^{(\alpha,\beta)}[-1,1]$ with $\alpha \geq \beta > -1$ and $\alpha + \beta \geq -1$ or $X = C[-1,1]$.

Other relations with the K-functions can be found in papers by Berens and Xu ([23] and [24]).

The results can be used to write the characterization in terms of the modulus w_s^J, but only for the case $s = 1$. For the un-weighted case ($\alpha = \beta = 0$) the main part modulus Ω_s can be replaced by the Ditzian-Totik modulus w_s^φ. Moreover, we can also use the τ modulus of Ivanov. Recall that Ivanov proved that the τ modulus is equivalent to the Ditzian-Totik one [173].

3.5 Felten modulus

The ideas presented in this section are due to Felten and are taken from [119] and [120]. For $x, h \in [-1,1]$ define

$$x \oplus h = x\sqrt{1 - h^2} + \sqrt{1 - x^2}\,h.$$

It can be proved that this is an inner operation on the unit interval. That is $\oplus : [-1,1]^2 \to [-1,1]$. Now define the differences as

$$(\Delta_h f)(x) = f(x \oplus h) - f(x)$$

and

$$\Delta_h^r f(x) = \Delta_h(\Delta_h^{r-1})f(x), \qquad r > 1.$$

Let us write X_∞ for $C[-1,1]$ and $X_p = L_p(dx/\varphi(x))$ for $1 \leq p < \infty$, where $f \in X_p$ means

$$\|f\|_{p,\varphi} = \left(\int_{-1}^{1} |f(x)|^p \frac{dx}{\sqrt{1-x^2}} \right)^{1/p} < \infty.$$

For $f \in X_p$ ($1 \leq p \leq \infty$) the modulus of order r is defined by

$$w_\varphi^r(f, t)_X = \sup_{|h| \leq t} \|\Delta_h^r f\|_p.$$

Theorem 3.5.1. *For $p \in [1, \infty]$, $r \in \mathbb{N}$, $\alpha \in (0, r)$ and $f \in X_p$ the following assertions are equivalent: (i) There exists a constant C such that, for all $n \in \mathbb{N}$, $E_n(f)_X \leq Cn^{-\alpha}$. (ii) There exists a constant K such that $w_\varphi^r(f, t) \leq Kt^\alpha$.*

The moduli $w_\varphi^r(f, t)$ are not well defined for un-weighted $L_p[-1,1]$ ($1 \leq p < \infty$). In particular, there are functions $f \in L_p[-1,1]$ for which the translations are not in $L_p[-1,1]$. On the other hand, the Felten modulus $w_\varphi^{2r}(f, t)_X$ of even order and the Butzer-Stens modulus [55] $w_r^T(f; \cos t)$ (3.15) are equivalent.

Chapter 4

Exact Estimates and Asymptotics

For $H \subset C[-1,1]$ we set

$$E_n(H) = \sup_{f \in H} E_n(f). \tag{4.1}$$

In this chapter we consider the global best approximation for some classes of functions.

We remark that, for trigonometric approximation, several exact results are known. Many of them were presented in a book by Korneichuk [210].

4.1 Asymptotics for $\mathrm{Lip}_1(M, [-1,1]$

The first estimate for some class of functions, as well as asymptotic, were given by Favard and Nikolskii.

Theorem 4.1.1 (Favard, [116]). *For each* $n \in \mathbb{N}$,

$$\frac{M}{n} < E_{n-1}(\mathrm{Lip}_1(M, [-1,1])) < \frac{M\pi}{2n}.$$

Given $A \subset [-1,1]$ and $f \in C[-1,1]$, set

$$E_n(f, A) = \inf_{p \in \mathbb{P}_n} \sup_{x \in A} |f(x) - P(x)|.$$

Assume $A = \{x_0, \ldots, x_n\}$, where $-1 \le x_0 < x_1 < \ldots < x_n \le 1$ and let $f_n : [-1,1] \to \mathbb{R}$ be a piece-wise linear continuous function such that $|f'_n(\theta)| = 1$ $(x_k < \theta < x_{k+1})$ and sign $f(x_k) = -$ sign $f_n(x_{k+1})$. Favard notice that this function is extremal in the following sense. If $f \in C[-1,1]$ satisfies a Lipschitz condition with constant 1, then

$$E_{n-1}(f, A) \leq E_{n-1}(f_n, A).$$

Nikolskii used this idea to obtain a strong version of the last theorem.

Theorem 4.1.2 (Nikolskii, [271]). *There exists a sequence $\{\varepsilon_n\}$ of positive numbers, $\varepsilon_n = \mathcal{O}(1/(n \log n))$ such that*

$$E_{n-1}(\mathrm{Lip}_1(M, [-1, 1])) = \frac{M\pi}{2n} - \varepsilon_n. \tag{4.2}$$

If $f \in \mathrm{Lip}_1[M, [-1, 1]]$, then

$$\limsup_{n \to \infty} n\, E_n(f) \leq \frac{M\pi}{2},$$

and there is a function in $\mathrm{Lip}_1(M, [-1, 1])$ for which equality holds.

Proof. We present the ideas if the proof for n even, $n = 2m$ (for n odd the proof is similar). Of course, we can consider only the case M_1.

Let $t_{n,k} = (2k - 1)\pi/(2n)$ be the zeros of the Chebyshev polynomial T_n, $1 \leq k \leq n$, and set $x_{n,k} = \cos(t_{n,k})$. Take $A = \{0\} \cup \{x_{n,k}, 1 \leq k \leq n\}$ and let f_n be the extremal function constructed above with respect to this set A. From Favard's theorem we know that

$$E_{n-1}(f_n) \leq E_{n-1}(\mathrm{Lip}_1(1, [-1, 1])) < \frac{\pi}{2n}.$$

Let $P_{n-1}(f_n)$ the polynomial of degree not greater than $n-1$ that interpolates f_n at the points $x_{n,k}$. That is

$$P_{n-1}(f, x) = \frac{1}{n} \cos(n \arccos x) \sum_{k=1}^{n} \frac{(-1)^{k-1} \sin t_{n,k}}{x - \cos t_{n,k}} f_n(\cos(t_{n,k})).$$

Notice that

$$P_{n-1}(f_n, 0) = \frac{1}{n} \sum_{k=1}^{n} (-1)^{k+m} \tan(t_{n,k}) f_n(\cos(t_{n,k}))$$

and the sign of the product $(-1)^{k+m} \tan(t_{n,k})$ does not change for $1 \leq k \leq m$ and for $m + 1 \leq k \leq n$. On the other hand, for $k = m$ and $k = m + 1$ the sign of $(-1)^{k+m} \tan(t_{n,k})$ is positive.

Take $Q_{n-1} \in \mathbb{P}_{n-1}$ such that $E_{n-1}(f, A) = \max\{| f(x) - Q_{n-1}(x) |, x \in A\}$. From the Chebyshev theorem we know that $f - Q_{n-1}$ alternates sign at the points

$x_{n,1}, \ldots, x_{n,m}, 0, x_{n,m+1}, \ldots, x_{n,n}$ and $\mid (f - Q_{n-1}(y) \mid = E_{n-1}(f, A)$, for $y \in A$. On the other hand

$$\mid f(0) - P_n(f, 0) \mid = \mid f(0) - Q_{n-1}(0) - P_n(f - Q_{n-1}, 0) \mid$$

$$= \left(1 + \frac{1}{n} \sum_{k=1}^{n} \mid \tan t_{n,k} \mid \right) E_{n-1}(f, A).$$

If we take into account that $f_n(0) = 0$ and $E_{n-1}(f, A) \leq E_{n-1}(f)$, the proof finishes by proving that

$$\mid P_{n-1}(f_n, 0) \mid \left(1 + \frac{1}{n} \sum_{k=1}^{n} \mid \tan t_{n,k} \mid \right)^{-1} = \frac{\pi}{2n} + \mathcal{O}\left(\frac{1}{n \ln n} \right). \qquad \square$$

4.2 Estimates for W^r

W^r is the family of all f such that $\mid f^{(r)}(x) \mid \leq 1$ almost everywhere. Moreover $W_p^r = \{ f : f^{(r-1)}$ is absolutely continuous on $[-1, 1]$ and $\| f^{(r)} \|_p \leq 1 \}$.

Theorem 4.2.1. *For all* $r \in \mathbb{N}$, $n \geq r - 1$ *and* $f \in C^r[-1, 1]$ *we have*

$$E_n(f) \leq \left(\frac{\pi}{2} \right)^r \frac{1}{(n+1)n \cdots (n-r+2)} \| f^{(r)} \|.$$

It follows from the last theorem that

Theorem 4.2.2. *For all* $r \in \mathbb{N}$ *and* $n \geq r - 1$ *we have*

$$E_n(W^r) \leq \left(\frac{\pi}{2} \right)^r \frac{1}{(n+1)n \cdots (n-r+2)}.$$

Theorem 4.2.3 (Bernstein, [29]). *One has*

$$\lim_{n \to \infty} n^r E_n(W^r) = K_r,$$

where K_r *is the Favard constant.*

Another proof of this equality was given by Fisher in [121]. Fisher recognized some properties of the solution of the extremal problem (4.1), with $H = W^r$.

Theorem 4.2.4 (Fisher, [121]). *Fix* $n > r$ *and a function* $f \in W^r$ *such that* $E_n(f) = E_n(W_r)$. *Then* $\mid f^{(n)}(x) \mid = 1$ *for all* $x \in [-1, 1]$ *and* f *has exactly* $n - r + 1$ *changes of sign in* $(-1, 1)$. *If* $n = r - 1$, *then* f *is a constant multiple of the Chebyshev polynomial.*

In particular

$$E_{n-1}(W_n) = \frac{2^{1-n}}{n!}.$$

In particular it follows from the last theorem that f is a *perfect spline* with exactly $n - r + 1$ knots on $(-1, 1)$, but we will not discuss here any property of splines.

Sinwel obtained some upper estimates for $E_n(W^r)$. The main idea was to reduce the problem to the trigonometric case (see also [316]).

Theorem 4.2.5 (Sinwel, [343]). *If $r \in \mathbb{N}$ and $n \geq r - 1$, then*

$$E_n(W^r) \leq \frac{K_r}{(n+1)\,n \cdots (n-r+2)}.$$

From Fisher's result we know that these estimates are not very good for small n.

4.3 Asymptotics for $C^{r,w}[-1,1]$

Other inequalities can be obtained if we assume some information related with the smoothness of the derivatives.

For $r \geq 0$ and a concave modulus of continuity w, let $C^{r,w}[-1,1]$ be the family of functions $f \in C^r[-1,1]$ such that $w(f^{(r)}, t) \leq w(t)$.

Theorem 4.3.1 ([29]). *Fix $r \in \mathbb{N}$, $\alpha \in (0,1)$ and set $w(t) = t^\alpha$. For each $n \in \mathbb{N}$, there exists a constant $C(r, \alpha, n)$ such that,*

$$E_n(C^{r,w}[-1,1]) \leq \frac{C(r, \alpha, n)}{n^{r+\alpha}},$$

and there exists a constant $C(r, \alpha)$ such that

$$\lim_{n \to \infty} C(r, \alpha, n) = C(r, \alpha).$$

For $r = 0$ a more exact result was obtained by Polovina in 1964.

Theorem 4.3.2 (Polovina, [283]). *Let w be a concave modulus of continuity and $H(w) = \{f \in C[-1,1] : w(f,t) \leq w(t)\}$. Then*

$$E_{n-1}(H(w)) = \frac{1}{2}w\left(\frac{\pi}{n}\right) - \varepsilon_n w\left(\frac{\pi}{n}\right)$$

with $\varepsilon_n = \mathcal{O}((\log n)^{-1})$.

In 1969 Polovina found an interesting lower bound.

Let $f_{n,0}$ be the odd $2\pi/n$-periodic function defined on $[0, \pi/n]$ by

$$f_{n,0}(x) = \begin{cases} w(2t)/2, & t \in [0, \pi/2n], \\ w(2\pi/n - 2t)/2, & t \in (\pi/2n, \pi/n] \end{cases}$$

and $f_{n,r}$ be the rth $2\pi/n$-periodic integral of $f_{n,0}$ with mean value on a period equal to zero.

Theorem 4.3.3 (Polovina, [285]). *Fix $r \in \mathbb{N}_0$ and a concave modulus of continuity w. For each $n \in \mathbb{N}$, one has*

$$E_n(C^{r,w}[-1,1]) \geq \|f_{n+1,r}\|_C (1 - \varepsilon_n),$$

where $\varepsilon_n = \mathcal{O}(1/\log n)$.

In 1980 Kofanov obtained the asymptotic.

Theorem 4.3.4 (Kofanov, [200]). *If $r \in \mathbb{N}_0$ and w is a concave modulus of continuity, then*

$$\lim_{n \to \infty} \frac{E_n(C^{r,w}[-1,1])}{\|f_{n,r}\|} = 1.$$

4.4 Estimates for integrable functions

For spaces of integrable functions the main results have been obtained for $L_1[-1,1]$.

The function $(1/\Gamma(r))(x - t)_+^{r-1}$ is known as the truncated power, here $\Gamma(r)$ stands for Euler's gamma-function. For algebraic approximation it has the same role as that of the Bernoulli kernels $D_r(t)$ in the theory of approximating 2π-periodic functions. One can reduce the problem of best approximation of some classes of functions to the problem of best approximation of truncated powers. For example, by the duality relation for the best approximation (see [261] and Theorem 1.2 in [203]),

$$E_n(V_1^r)_1 = \sup_{a \in [-1,1]} E_n((x - a)_+^{r-1})_1,$$

where V_1^r is the class all functions f which can be represented in the form

$$f(x) = \frac{1}{\Gamma(r)} \int_{-1}^{1} (x - t)_+^{r-1} \phi(t) dt,$$

where $\phi \in L_1[-1,1]$, $\|\phi\|_1 \leq 1$.

Set $s_n(t) = \operatorname{sign} \sin(n+2) \arccos t$ and define

$$s_{n,r}(x) = \frac{1}{(r-1)!} \int_{-1}^{1} (x - t)_+^{r-1} s_n(t) dt.$$

Theorem 4.4.1 (Kofanov, [201] and [202]). *If $r \in \mathbb{N}$ and $n \geq r - 1$, then*

$$E_n(W_1^r)_1 = \|s_{n,r}\|_\infty. \tag{4.3}$$

Moreover, If $r \geq 2$ and $f^{(r-1)}$ is absolutely continuous and $f^{(r)} \in L_1$, then for $n > r - 1$,

$$E_n(f)_1 \leq \|s_{n,r}\|_\infty E_{n-r}(f^{(r)})_1.$$

A similar result was presented in [203], but for the class V_1^r, with a real $r \geq 1$. In this case we also have the equality (4.3), but for $n \geq [r] - 1$.

Theorem 4.4.2 (Motornaya, [263]). *For $r \in \mathbb{N}$, one has*

$$E_n((x-a)_+^{r-1})_1 = \frac{K_k}{(n+1)^r} \left(\sqrt{1-a^2}\right)^r (r-1)! + \mathcal{O}\left(\frac{(\sqrt{1-a^2})^{r-1}}{(n+1)^{r+1}}\right),$$

where $a \in (-1,1)$ and $n \geq r - 1$.

This result was used to obtain the following estimate.

Theorem 4.4.3 (Motornaya, [264]). *For $r \in \mathbb{N}$, one has*

$$E_n(W_\infty^r)_1 = \frac{2}{\pi} B\left(\frac{1}{2}, \frac{r}{2}+1\right) \frac{K_{r+1}}{(n+1)^r} + o\left(\frac{1}{n^r}\right),$$

where $B(x,y)$ is the Euler integral of the first kind and K_r are the Favard constants.

The best approximation of the classes W_p^r by algebraic polynomials P_n in the L_q norm is defined by:

$$E_n(W_p^r)_q = \sup_{f \in W_p^r} \inf_{u \in P_n} \|f - u\|_q, \quad 1 \leq p, q \leq \infty.$$

Motornyi and Motornaya had obtained some asymptotic in L_1 norm. In [259] some estimates were announced without proof. For instance,

$$E_n(W_p^r)_1 = \left(\frac{1}{2\pi} \int_{-1}^1 (1-t^2)^{rq/2} dt\right)^{1/q} \|\varphi_{n,r}\|_q + o(1/n^r),$$

where $1/p + 1/q = 1$ and $\varphi_{n,r}$ is the r-periodic integral of the function $\operatorname{sign} \sin(n+1)t$, whose mean value on the period is equal to zero. In [260] they considered the class $W^r H^\alpha$, $r = 0, 1, \ldots$, and $\alpha \in (0,1]$, ($f^{(r)} \in \operatorname{Lip}_\alpha[-1,1]$). For this class they obtained the asymptotic

$$E_n(W^r H^\alpha)_1 = \frac{1}{2\pi} \int_{-1}^1 (1-t^2)^{(r+\alpha)/2} dt \, \|f_{n,r,\alpha}\|_1 + o\left(\frac{1}{n^{r+\alpha}}\right),$$

where $f_{n,r,\alpha}$ is the rth periodic integral of the $2\pi/n$-periodic odd function

$$f_{n,0,\alpha}(t) = \begin{array}{ll} 2^{\alpha-1} t^\alpha, & 0 \leq t \leq \pi/(2n), \\ 2^{\alpha-1}(\pi/n - t)^\alpha, & \pi/(2n) \leq t \leq \pi/n. \end{array}$$

In [262] a review of the approximation of certain functions and classes of functions by algebraic polynomials in the spaces C and L_1 is presented.

For Jacobi weights, Rafalson provided some estimates for weighted approximation in the spaces $L_{p,\alpha,\beta}$.

Set

$$\psi_r^{[k]}(f,x) = \begin{cases} \left((1-x)^{r+\alpha}(1+x)^{r+\beta}f^{(r)}(x)\right)^{(r)}, & |x| < 1, \\ 0, & |x| = 1 \end{cases}$$

and

$$\Omega_r = \{f : f \in C^{(2r-1)}(-1,1), \psi_r^{[k]}(f,x) \in AC[-1,1], 0 \le k \le r-1\}.$$

For $f \in \Omega_r$, define

$$D_r(f,x) = \frac{1}{(1-x)^\alpha(1+x)^\beta}\left((1-x)^{r+\alpha}(1+x)^{r+\beta}f^{(r)}(x)\right)^{(r)}.$$

Finally, for $t \in (-1,1)$, define

$$\Phi_r(t) = \frac{(-1)^r}{2^{\alpha+\beta+r}}\frac{\Gamma(r+\alpha+\beta+1)}{\Gamma^2(r)\Gamma(\alpha+1)\Gamma(\beta+r)}$$
$$\times \int_{-1}^t \frac{(t-z)^r}{(1-z)^{1+\alpha}(1+z)^{r+\beta}} \int_{-1}^z (1-u)^\alpha(1+u)^{\beta+r-1}du\,dz.$$

Theorem 4.4.4 (Rafalson, [312]). *If $r, n+1 \in \mathbb{N}$, $n \ge r-1$ such that $r > \alpha+1$ and $q \in [1, \infty]$, or $r = \alpha+1$, $q \in [1, \infty)$, or $r < \alpha+1$, and $q \in [1, (1+\alpha)/(1+\alpha-r))$, then*

$$\sup\left\{\frac{E_n(f)_{p,\,\alpha,\,\beta}}{E_n(D_r(f))_{1,\,\alpha,\,\beta}}, f \in \Omega_r, E_n(D_r(f))_{1,\,\alpha,\,\beta} \neq 0\right\} = E_n(\Phi_r)_{q,\,\alpha,\,\beta}.$$

There are other papers of Rafalson related with this kind of problem.

4.5 Pointwise asymptotics

Usually, the construction of a linear method to approximate continuous functions by means of algebraic polynomials is done with the help of Chebyshev polynomials. Consider the orthonormal polynomials

$$\widetilde{T}_n(x) = \sqrt{\frac{2}{\pi}}\cos(n\arccos x)), n \in \mathbb{N}$$

and $\widetilde{T}_0(x) = \sqrt{1/\pi}$. For $f \in C[-1,1]$ the Fourier-Chebyshev coefficients are given by

$$c_k(f) = \int_{-1}^1 \frac{f(t)\widetilde{T}_k(t)}{\sqrt{1-t^2}}dt.$$

For a matrix $\Lambda = \{\lambda_{k,n}\}$, $k, n \in \mathbb{N}_0$, the linear operator U_n is defined by

$$U_n(f, x) = \sum_{k=}^{n} \lambda_{k,n} c_k(f)\widetilde{T}_k(x).$$

In the case when U_n corresponds to the arithmetical means $\sigma_{n,n}$ of the Fourier-Chebyshev series ($\lambda_{k,n} = (n-k+1)/(n+1)$ and $0 < \alpha \le 1$, Ganzburg and Timan obtained an asymptotic related with these operators.

Theorem 4.5.1 (Ganzburg and Timan, [132]). *For $n \in \mathbb{N}$, $0 < \alpha < 1$ and $x \in [-1, 1]$, set*

$$E_n^{(\alpha)}(x) = \sup_{f \in \mathrm{Lip}_\alpha(1, [-1,1])} | f(x) - \sigma_{n,n}(f, x) | .$$

Then

$$E_n^{(\alpha)}(x) = \frac{2\Gamma(\alpha)}{\pi(1-\alpha)} \sin \frac{\alpha\pi}{2} \left(\frac{\sqrt{1-x^2}}{n} \right)^\alpha + o \left[\left(\frac{\sqrt{1-x^2}}{n} \right)^\alpha \right] + \delta_n^\alpha(x),$$

where

$$\delta_n^\alpha(x) = \begin{cases} \mathcal{O}(| x |^\alpha / n^{2\alpha}), & \text{if } 0 < \alpha < 1/2, \\ \mathcal{O}(| x |^\alpha / n^\alpha), & \text{if } \alpha > 1/2, \\ \mathcal{O}(\sqrt{| x |} \log n / n), & \text{if } \alpha = 1/2. \end{cases}$$

For Jackson-Timan-type results some good constants were obtained by Runck and Sinwel in 1980.

Theorem 4.5.2 (Runck and Sinwel, [316]). *For $r \in \mathbb{N}$, $f \in W^r$ and $n > 2r$ there exists a polynomial $P_n \in \mathbb{P}_n$ such that*

$$| f(x) - P_n(x) | \le \frac{K_r}{(n-2)(n-4)\cdots(n-2r)} \left(\sqrt{1-x^2} + \frac{2r}{n} | x | \right)^r.$$

For all $f \in W_1$ and $n > 1$, there exists a polynomial $P_n \in \mathbb{P}_n$ such that

$$| f(x) - P_n(x) | \le \tan \frac{\pi}{2n} \left(\sqrt{1-x^2} + \frac{3}{n} | x | \right).$$

Moreover, for all $n > 1$ there exists $f \in W_1$, such that for all $P_n \in \mathbb{P}_n$ there exists an $x \in [-1, 1]$, so that

$$| f(x) - P_n(x) | \ge \tan \frac{\pi}{2n} \sqrt{1-x^2}.$$

In a series of papers ([211], [212] and [213]) Korneichuk and Polovina improved the asymptotic given in (2.3) and (2.7). Typical results are the following.

Theorem 4.5.3 (Korneichuk and Polovina, [212]). *Fix $r \geq 0$ and $0 < \alpha < 1$. If $f \in C^r[-1,1]$ and $\omega(f^{(r)}, t) \leq Kt^\alpha$, there exists a sequence $\{P_n\}$ ($P_n \in \mathbb{P}_n$) such that, for $x \in [-1,1]$,*

$$| f(x) - P_{n-1}(f,x) | \leq \frac{K}{2} \left(\frac{\pi}{2} \sqrt{1-x^2} \right)^\alpha + \mathcal{O}(n^{-3\alpha/2}).$$

The case $\alpha = 1$ was studied by Nikolskii (see Theorem 4.1.2).

Theorem 4.5.4 (Korneichuk and Polovina, [213]). *Let w be a modulus of continuity. For any function $f \in H_w(1, [-1,1])$ there is a sequence of algebraic polynomials $\{P_n(f)\}$ ($P_n \in \mathbb{P}_n$) such that*

$$| f(x) - P_n(f,x) | \leq Cw \left(\frac{\pi \sqrt{1-x^2}}{n+1} \right) + o \left(w \left(\frac{1}{n+1} \right) \right)$$

where C can be taken as 1. Moreover, if w is a concave modulus of continuity, C can be taken as $1/2$.

In the papers of Korniechuk-Polovina and Ligun (see Theorem 2.3.4) the generalization of the Nikolskii theorem was accompanied by improving the remainder. In the proof they used the intermediate-approximation method to obtain exact estimates for the deviation of best approximations to the class of periodic functions. The construction is carried out with a nonlinear operator. In particular, from the results of Korneichuk and Polovina [211] it follows that for all $f \in W_2[-1,1]$ there exists a sequence of polynomials $\{P_n(f)\}$ satisfying the inequality

$$| f(x) - P_n(f,x) | \leq K_2 \frac{1-x^2}{(n+1)^2} + o \left(\frac{1}{(n+1)^2} \right).$$

Trigub considered the problem of the leading term in the corresponding pointwise inequality concerning approximation of functions in $W^r[-1,1]$.

Theorem 4.5.5 (Trigub, [390]). *For each $r \in \mathbb{N}$ there exists a constant $\gamma = \gamma(r)$ such that, for all $f \in W^r[-1,1]$ there exists a sequence $\{P_n(f)\}$ ($P_n(f) \in \mathbb{P}_n$, $n \geq r-1$) such that*

$$| f(x) - P_n(f,x) | \leq K_r \left(\frac{\sqrt{1-x^2}}{n+1} \right)^r + \gamma \frac{(\sqrt{1-x^2})^{r-1}}{(n+1)^{r+1}}.$$

Here it is necessary that $\gamma(r) \leq ce^r$, where c is a positive constant.

Tribug also studied the problem in weighted spaces. Let $V_1[-1,1]$ be a class of functions with total variation of the derivative $f^{(r-1)}$ not greater than one.

Theorem 4.5.6 ([390]). *For each $r \in \mathbb{N}$ and for all $f \in V_r[-1,1]$ there exists a sequence $\{P_n(f)\}$ ($P_n(f) \in \mathbb{P}_n$, $n \geq r-1$) that satisfies the inequality*

$$\int_{-1}^{1} \frac{| f(x) - P_n(f,x) |}{(1-x^2)^{r/2}} dx \leq \frac{K_r}{n^r} + o(n^{-r}).$$

For $r = 1$, there is no remainder term.

In the first theorem of Trigub for point-wise approximations the polynomial operator $P_n(f)$ is nonlinear and for weighted approximations in $L_1[-1,1]$ it is linear. He noticed that, if we multiply by two the right-hand side in the first theorem, then the result can be given with a linear operator. Trigub noticed that his ideas can be used to extend the results by means of K-functionals: for all $f \in C^r[-1,1]$, there exists a sequence of polynomials $\{P_n\}$ such that

$$| f(x) - P_n(x) | \leq \gamma \gamma_k \left(\frac{\sqrt{1-x^2}}{n} \right)^r \omega_k \left(f^{(r)}, \frac{\pi}{2} \frac{\sqrt{1-x^2}}{n} \right).$$

In this way he recovered some known results (see Theorem 5.6.5), but the difference of his proof (which can be also applied to approximation of derivatives) is that the results are deduced directly from the periodic case (without constructing especial integral operators). Trigub asserted that by using the approximate characterization of W^r in the periodic case, one can get an approximate characterization of W^r on a segment (see [389]).

Let us present a group of ideas of Motornyi taken from [257]. Some of the proof of the asymptotic (as the one given by Korneichuk and Polovina, [213]) is based on the method of intermediate approximation. For any function f with convex modulus of continuity $w(f,t)$, one can construct a sequence of broken lines $\psi_n(x)$ possessing the following properties:

(i) If $\psi_n'(x)$ exists, then

$$| \psi_n'(x) | \leq w' \left(\frac{\pi}{n} \sqrt{1-x^2} \right) = K_n(x), \quad n = 2,3,\ldots, \quad | x | \leq 1.$$

(ii) The inequality

$$| f(x) - \psi_n(x) | \leq \frac{1}{2} \max_{0 \leq t \leq 2} \{w(t) - K_n(x)t\} + o\left(w\left(\frac{1}{n} \right) \right)$$

holds uniformly with respect to $x \in [-1,1]$ as $n \to \infty$. Let $x_0 = 0$ and

$$x_k = x_{k-1} + \frac{a}{n} \sqrt{1 - x_{k-1}^2}, \quad n \geq 5$$

be points of the segment $[0,1]$. Here, $a \in [1,\pi]$ is a constant. Let x_{N-1} denote the greatest point for which $x_{N-1} \leq \overline{x}$, where the number $\overline{x} < 1$ is such that $\overline{x} + a\sqrt{1-\overline{x}^2}/n = 1$. If $x_{N-1} = \overline{x}$, then we have $x_N = 1$, and if $x_{N-1} < \overline{x}$, then, by definition, we assume that $x_N = 1$. We set

$$E_k = [-x_{k+1}, -x_k] \cup [x_k, x_{k+1}], \quad k = 0, 1, \ldots, N-1.$$

Theorem 4.5.7 (Motornyi, [257]). *Suppose that $w(t)$ is a convex modulus of continuity. Then, for any function $f \in H_w$ and any number $a \in [1,\pi]$, there exists* a sequence of absolutely continuous functions $\{\psi_{n,a}(f;x)\}$ such that the following assertions are true:

(i) *The inequality*

$$|\psi'_{n,a}(x)| \leq M_{k+1}, \quad x \in E_k, \quad k = 0, 1, \ldots, N_1,$$

holds almost everywhere;

(ii) $|f(x) - \psi_{n,a}(x)| \leq \Delta_k$, $x \in E_k$, $k = 0, 1, \ldots, N_1$, *where*

$$M_k = w' \frac{a\sqrt{1 - x_{k-1}^2}}{n}$$

and

$$\Delta_k = \frac{1}{2}\left(w\left(\frac{a\sqrt{1 - x_{k-1}^2}}{n} - M_k \frac{a\sqrt{1 - x_{k-1}^2}}{n}\right)\right).$$

This theorem is a generalization of the result of Korneichuk and Polovina on the approximation of functions from the class H_w by absolutely continuous functions with variable smoothness.

Theorem 4.5.8 (Motornyi, [257]). *Suppose that $w(t)$ is a convex modulus of continuity such that the function $tw'(t)$ does not decrease. Then, for any function $f \in W^r H_w$, there exists a sequence of algebraic polynomials $Q_{n,r}(f,x)$ of degree $n = r, r+1$, ($n \geq 2$ for $r = 0$) such that*

$$|f(x) - Q_{n,r}(f,x)| \leq \frac{K_r}{2}\left(\frac{\sqrt{1 - x^2}}{n}\right)^r w\left(\frac{2K_{r+1}\sqrt{1 - x^2}}{K_r n}\right)$$
$$+ \frac{C_r}{n^{r+1}}\left(\sqrt{1 - x^2} + \frac{1}{n}\right)^{r-1} w\left(\frac{\sqrt{1 - x^2}}{n} + \frac{1}{n^2}\right) \log n,$$

where K_r is the Favard constant and the quantity C_r depends only on r.

Classes of functions which are singular integrals of bounded functions were considered by Motornyi in [258].

Chapter 5

Construction of Special Operators

In the previous chapters we paid attention to theorems related with the existence of sequences of polynomials satisfying certain conditions. In applications this kind of results are not important. What we need is a way to obtain the polynomials with the desired properties. Many authors have constructed different sequences.

From the point of view of application the sequences should satisfy some of the following conditions:

1) We need the precise form of the polynomials.
2) The construction should be useful for numerical computation.
3) A clear form of measuring the error is needed. Many proofs have been presented in such a way that it is very difficult to give a good estimate of the constants. In such cases we do not know the minimal degree of the polynomial which will be used to obtain a fixed error. On the other hand, several theorems are presented in terms of the best approximation but, as it is known, there is no easy way to find the best approximation for a given function.
4) In some cases we can not use all the values of the function. In several practical problems, we only have a discrete set of data and we wish to reconstruct the function. But many useful approximation processes are constructed by means of integrals, for instance, convolution with some kernels. One can consider the problem of the best selection of a collection of nodes, but usually the data are given only on equidistant nodes.
5) Sometimes we need a certain subspace Q to be invariant. Thus, the operator is a linear projection. Usually Q is a family of polynomials, $Q = \mathbb{P}_n$. We want an operator $L_n : C[-1,1] \to C[-1,1]$ such that $L_n(p) = p$, for each $p \in \mathbb{P}_n$. It is a strong restriction. In fact, for $f \in C[-1,1]$ one has

$$\|f - L_n f\| = \|(I - L_n)(f - p)\| \le \|I - L_n\| \, \|f - p\|$$

for each $p \in \mathbb{P}_n$. Thus

$$\|f - L_n f\| \leq \|I - L_n\| E_n(f) \leq (1 + \|L_n\|) E_n(f)$$

and it can be proved that for any projection $L_n : C[-1,1] \to \mathbb{P}_n$, one has $\|L_n\| \to \infty$.

6) We also considered the problem of constructions of linear operators of minimal degree compared to the number of points of interpolation. It will be better if, at the same time, they realize the Teliakovskii-Gopengauz estimate.

In order to improve the results, different methods have been employed:

1) Some authors have used known facts related with trigonometric approximation (for instance, convolution with positive even kernel).

2) We can construct a sequence of algebraic polynomials by means of a suitable interpolation process. It is very useful when we have only a finite set of data but, as we remarked above, the sequence of the norm of the associated operators may be unbounded. In particular, Lagrange interpolation at equidistant nodes has a very bad behavior. On the other hand, some modification (such as the Hermite-Fejér interpolation) give a good rate of convergence for certain classes but these processes are saturated. That is, the order of convergence cannot be improved upon beyond a certain limit.

Other general criteria have also been considered. For instance, some people prefer processes constructed by means of linear operators and the best ones are those which are uniformly bounded. This kind of process are more convenient for applications, because they are easier to handle.

In this chapter we present different methods which have been proposed. In the first section, we consider estimation in norm. In the second and third section we analyze estimates in the form of Timan-type and Teliakovskii-Gonpengauz-type inequalities. Since there are many papers devoted to interpolation processes of Bernstein type, they will be analyzed in a separate section.

Following Freud and Sharma we say that an approximation process is of Timan type if the rate of convergence of the function $\Delta_n(x)$ can be given. A process is said to be weakly interpolatory, if it is uniquely determined by the values of the given function on a finite set.

In this chapter we will use several times the following operator: if $f : [a,b] \to \mathbb{R}$, we set

$$L(f, x) = \frac{x - a}{b - a} f(b) + \frac{b - x}{b - a} f(a). \tag{5.1}$$

From Cao and Gonska [69] we use the following terminology.

Definition 5.0.1. Given $r \in \mathbb{N}$ and $s \in \mathbb{Z}$, a sequence of linear operators $L_n : C[-1,1] \to \mathbb{P}_{rn+s}$ is said to be of DeVore-Gopengauz type, if there exists a constant C such that, for all $f \in C[-1,1]$, $n \in \mathbb{N}$ and $x \in [-1,1]$,

$$| f(x) - L_n(f, x) | \leq C \omega_2 \left(f, \frac{\sqrt{1 - x^2}}{n} \right).$$

5.1 Estimates in norm

At the beginning of the theory, theorems for algebraic approximation were obtained from the trigonometric approach, which uses the Jackson kernels. Thus, a natural problem was to obtain similar results by means of polynomial operators.

Recall that, given points $-1 \leq x_1 < \cdots < x_n \leq 1$ and $f : [-1,1] \to \mathbb{R}$ the Lagrange interpolation operator is defined by

$$L_n(f,x) = \sum_{k=1}^{n} l_{k,n}(x) f(x_k), \tag{5.2}$$

where $w(x) = (x - x_1) \cdots (x - x_n)$ and

$$l_{k,n}(x) = \frac{w(x)}{w'(x)(x - x_k)}, \qquad k = 1, \ldots, n,$$

are the fundamental polynomials of the Lagrange interpolation.

Let us recall some facts. The Chebyshev polynomials (of the first kind) are defined by

$$T_n(x) = \cos(n \arccos x).$$

The zeros of T_n are

$$x_{k,n} = \cos \frac{(2k-1)\pi}{2n} \qquad (k = 1, 2, \ldots, n). \tag{5.3}$$

The fundamental Lagrange interpolation polynomials relative to these nodes are

$$l_{k,n}(x) = \frac{(-1)^{k+1} \sqrt{1 - x_{k,n}^2}}{n} \frac{T_n(x)}{x - x_{k,n}}, \qquad k = 1, 2, \ldots, n. \tag{5.4}$$

It is known that there exist functions $f \in [-1,1]$ for which the sequence of polynomials obtained from the Lagrange interpolation formula diverges for all points of the interval $[-1,1]$ (see [153]). Grünwald showed that modification of the Lagrange operators with Chebyshev nodes can be used as an approximation process. The result is analogous with the theorem of Rogosinski in the theory of Fourier series.

Theorem 5.1.1 (Grünwald, [154]). *Let $\{L_n\}$ be the sequence (5.2) constructed with the Chebyshev nodes. For each $f \in [-1,1]$ one has*

$$\lim_{n \to \infty} \frac{1}{2} [L_n(f, \theta - \pi/2n) + L_n(f, \theta + \pi/2n)] = f(x), \qquad x = \cos \theta,$$

and the convergence is uniform on the whole interval.

Another simple construction was given by Freud in 1963. He also used the Chebyshev nodes (5.3).

Theorem 5.1.2 (Freud, [124]). *If for $n \in \mathbb{N}$ and $f \in C[-1, 1]$ we set*

$$F_n(f, x) = f(0) + \sum_{k=1}^{n} l_{k,n}(x)[f(x_{k,n}) - f(0)],$$

then, there exists a constant C such that, for $n \in \mathbb{N}$, $f \in C[-1, 1]$ and $x \in [-1/2, 1/2]$, one has

$$| f(x) - F_n(f, x) | \leq C\omega\left(f, \frac{1}{4n}\right).$$

Notice that $F_n(f) \in \Pi_{4n-3}$ and Freud only proved uniform convergence on the interval $[-1/2, 1/2]$.

Freud's work motivated a series of papers exhibiting constructions of a similar character. For instance, Sallay [322] obtained an analogous result, but with interpolation at the zeros of orthogonal polynomials with respect to a weight function $w \in \mathrm{Lip}_1[-1, 1]$ which is positive on $[-1, 1]$.

In 1967 Saxena modified Freud's ideas and used interpolation at the zeros of Chebyshev polynomials of the second kind [323]. These polynomials are defined by

$$U_n(x) = \frac{\sin((n + 1)\theta)}{\sin \theta}, \qquad \cos \theta = x. \tag{5.5}$$

With the new construction Saxena was able to obtain convergence on the whole interval $[-1, 1]$.

In the same year Vértesi noticed that the Saxena ideas could be modified to use Chebyshev polynomials T_n instead of U_n. Of course, more complicated operators appeared. Set

$$v_{k,n}(x) = 1 - \frac{x_{k,n}}{1 - x_{k,n}^2}(x - x_{k,n}), \qquad \psi_n(u, v) = \frac{2}{n}\sum_{r=1}^{n-1} T_r'(u)T_r(v),$$

and

$$\varphi_{k,n}(x) = v_{k,n}(x)l_{k,n}^4(x) + 2(x - x_{k,n})l_{k,n}^3(x)\psi_n(x_{k,n}, x),$$

where $l_{k,n}$ is given by (5.4).

Theorem 5.1.3 (Vértesi, [397]). *If for $n \in \mathbb{N}$ and $f \in C[-1, 1]$ we set*

$$J_n(f, x) = L(f, x) + \sum_{k=1}^{n} (f(x_{kn}) - L(f, x))\, \varphi_{kn}(x), \tag{5.6}$$

where $L(f, x)$ is defined by (5.1), then

$$| f(x) - J_n(f, x) | \leq 512\,\omega\left(f, \frac{1}{4n}\right), \qquad x \in [-1, 1].$$

In this case we have uniform convergence on the whole interval and we have also a precise constant.

In 1971 Mathur [249] presented estimates by interpolating in the zeros of the Jacobi polynomials $P_n^{(-1/2,1/2)}$. Srivastava gave some modifications in [345].

For approximation in norm, Varma showed how to construct sequences of polynomials $\{P_n\}$, $P_n \leq \Pi_{2n-1}$, which interpolates f at the zeros of the nth Chebyshev polynomials and for which $\|f - P_n\| \leq C_k \omega_k(f, 1/n)$. The construction is based in a modification of the classical Hermite-Fejér interpolation polynomial on the Chebyshev nodes.

Define

$$C_0(f) = \frac{1}{n} \sum_{k=1}^{n} f(x_{k,n}), \qquad C_j(f) = \frac{2}{n} \sum_{k=1}^{n} f(x_{k,n}) T_j(x_{k,n}),$$

for $j = 1, \ldots, 2n - 1$. For a fixed sequence $\{\alpha_{j,n}\}$ set

$$R_n(f, x) = \sum_{j=0}^{2n-1} C_j(f) \alpha_{j,n} T_j(x). \tag{5.7}$$

The motivation for this definition comes from the following: if

$$\alpha_{j,n} = \frac{2n - j}{2n},$$

the $R_n(f)$ agrees with the Hermite-Fejér interpolation polynomial.

Theorem 5.1.4 (Varma, [392]). *Given fixed $m \in \mathbb{N}$, for each $n \in \mathbb{N}$ consider a numerical sequence $\{\alpha_{j,n}\}$ such that*

(i) $\quad \alpha_{0,m} = 1, \quad \alpha_{j,m} + \alpha_{2n-j,m} = 1, \quad j = 1, \ldots, n, \quad \alpha_{j,m} = 0 \; (j > 2n),$

(ii) $\qquad\qquad\qquad 1 - \alpha_{1,m} = \mathcal{O}(1/n^m)$

(iii) $\qquad | \alpha_{j+1,m} - 2\alpha_{j,m} + \alpha_{j-1,m} | = \mathcal{O}(1/n^2), \quad j = 1, \ldots, 2n - 1$

(iv) $\qquad\quad | \mu_{j+1,m} - \mu_{j,m} | = \mathcal{O}(1/n^{m+1}), \quad j = 1, \ldots, 2n - 1$

(v) $\qquad | \mu_{j+1,m} - 2\mu_{j,m} + \mu_{j-1,m} | = \mathcal{O}(1/n^{m+2}), \quad j = 1, \ldots, 2n - 1,$

where
$$\mu_{j,m} = (1 - \alpha_{j,m})/j^m, \quad j = 1, \ldots, 2n,$$

and $\mu_{j,m} = 0$, for $j = 0$.
If R_n is defined by (5.7), then $R_n(1, x) = 1$ and for each $f \in C[-1, 1]$,

$$R_n(f, x_{k,n}) = f(x_{k,n}), \quad k = 1, \ldots, n.$$

Moreover, there exists a constant C_m such that, for each $f \in C[-1, 1]$ and $n \in \mathbb{N}$,

$$\|f - R_n(f)\| \leq C_m \omega_{m-1}(f, 1/n).$$

Proof. We will present the main ideas of the proof. Taking into account that, for $1 \leq j \leq n$, $T_{2n-j}(x_{i,n}) = -T_j(x_{i,n})$, one has $C_{2n-j}(f) = -C_j(f)$. Thus, from (i) and the definition of $C_j(f)$ we obtain

$$R_n(f, x_{i,n}) = C_0(f) + \sum_{j=1}^{n-1} C_j(f) T_j(x_{i,n}) = f(x_{i,j}), \qquad 1 \leq i \leq n.$$

The identity $R_n(1, x) = 1$ follows directly from the definition of $C_j(f)$. Notice that $C_j(1) = 0$, for $1 \leq j \leq 2n - 1$.

It can be proved that there exists a constant L such that $\|R_n(f)\| \leq L\|f\|$. In fact, we can write

$$R_n(f, x) = \sum_{k=1}^{n} f(x_{k,n}) P_{k,n}(x)$$

where

$$P_{k,n}(x) = \frac{1}{n}\left(1 + 2\sum_{j=1}^{2n-1} \alpha_{j,m} T_j(x_{k,n}) T_j(x)\right).$$

Hence, we only need to estimate $\sum_{k=1}^{n} | P_{k,n}(x) |$.

Set $t_1(s) \equiv 1$,

$$t_j(s) = 1 + \frac{2}{j}\sum_{i=1}^{j-1}(j - i)\cos(is), \qquad j \geq 2.$$

and

$$\tau_{j,k}(s) = \frac{1}{2}\left(t_j(s + \theta_{k,n}) + t_j(s - \theta_{k,n})\right)$$

where $\theta_{k,n} = (2k - 1)\pi/(2n)$. It can be proved that, for $j \geq 1$,

$$\sum_{k=1}^{n} | \tau_{j,k}(s) | = n \tag{5.8}$$

and

$$(j + 1)\tau_{j+1,k}(s) - 2j\tau_{j,k}(s) + (j - 1)\tau_{j-1,k}(s) = 2\cos(js)\cos(j\theta_{k,n}).$$

From the last identity we obtain the representation

$$P_{k,n}(x) = \frac{1}{n}\sum_{i=1}^{2n-1}(\alpha_{i+1,m} - 2\alpha_{i,m} + \alpha_{i-1,m})\tau_{i,k}(s) + \tau_{2n,k}(s)\alpha_{2n-1,n}.$$

Finally, from conditions (ii) and (iii) and (5.8), we obtain a constant L such that $\sum_{k=1}^{n} | P_{k,n}(x) | \leq L$.

The estimate $\|R_n(f)\| \leq L\|f\|$ is sufficient to prove uniform convergence. We omit the proof of the estimate in terms of the modulus of continuity. $\qquad\square$

An example of sequences satisfying all the conditions stated above is given by

$$\alpha_{j,m} = \frac{(2n-j)^m}{(2n-j)^m + j^m}, \qquad j = 0, \ldots, 2n-1,$$

and $\alpha_{2n,m} = 0$.

5.2 Timan-type estimates

The kernels constructed by Dzyadyk ([109] and [110]) allowed him to give a new proof of Timan's theorem (see Theorem 2.5.1). But the polynomials obtained by this way cannot always be constructed effectively, since their coefficients are computed in terms of integrals of the function to be approximated.

Freud and Vértesi noticed that the construction given by Vértesi in [397] could be used to provide a new proof of the Timan result.

Theorem 5.2.1 (Freud and Vértesi, [128]). *For each n, let J_n be defined by (5.6). Then $J_n(C[-1,1]) \subset \Pi_{4n-2}$ and there exists a constant C such that, for $f \in C[-1,1]$ and $x \in [-1,1]$, one has*

$$\mid f(x) - J_n(f,x) \mid \le C\left(\omega\left(f, \frac{\sqrt{1-x^2}}{4n}\right) + \omega\left(f, \frac{1}{(4n)^2}\right)\right).$$

Another construction was given by Kis and Vértesi [199] (see 85.37).

In 1968, Stepanets and Poliakov [351] gave a new proof of Theorem 2.4.1. They used polynomials whose coefficients are expressed in terms of the values of the functions and its derivatives (if they exist) at a finite system of points. In the construction they used the Dzyadyk kernel D_{nk} given in (2.12). Since the construction is a little complicated it will not be included here.

In 1974, Mills and Varma used a sequence obtained by means of Lagrange interpolation, but it was combined with the construction of Grünwald.

Set

$$l_{k,n}(x) = \frac{(-1)^{k+1}\cos n\theta \cos\theta_{n,k}}{n(\cos\theta - \cos\theta_{k,n})},$$

where, as usual, $x = \cos\theta$ and $x_{k,n}$ are the zeros of the Chebyshev polynomials. Now define

$$G_n(f,\theta) = \frac{1}{2}\sum_{k=1}^{n}\left[l_{k,n}\left(\theta + \frac{\pi}{2n}\right) + l_{k,n}\left(\theta - \frac{\pi}{2n}\right)\right]f(x_{k,n}).$$

Theorem 5.2.2 (Mills and Varma, [251]). *If $f \in C[-1,1]$, then*

$$\mid f(\cos\theta) - G_n(\theta) \mid \le C\left(\omega\left(f, \frac{\sqrt{1-x^2}}{n}\right) + \omega\left(f, \frac{1}{n^2}\right)\right).$$

In this result, $G_n(C[-1,1]) \subset \mathbb{P}_{n-1}$ and only n values of f are needed.

Almost all the interpolatory result presented above used the zeros of the Chebyshev polynomials.

In 1974 and 1977, Freud and Sharma constructed operators based on general Jacobi nodes, and also succeeded in decreasing the degree of the polynomial to $n(1+\varepsilon)$, for an arbitrary $\varepsilon > 0$.

Let $\{x_{k,n}\}$ be the zeros of $P_n^{(\alpha,\beta)}$ and let

$$l_{n,k}(x) = \frac{P_n^{(\alpha,\beta)}(x)}{(x - x_{kx})(P_n^{(\alpha,\beta)})'(x_{n,k})}$$

be the fundamental polynomials of Lagrange interpolation at these nodes.

Let $r \geq 2$ be an integer and fix $\rho \in (0, 1/2r)$. For a given $n \in \mathbb{N}$ we set $m = m(n) = [n\rho]$ and define

$$\Phi_n(x, y) = \frac{1}{m}\left(1 + 2\sum_{j=1}^{m} T_j(x)T_j(y)\right),$$

where T_j is the Chebyshev polynomial. In terms of the Lagrange basis one has

$$\Phi_n^{2r}(x, y) = \sum_{k=1}^{n} \Phi_n^{2r}(x_{kn}, y)l_{kn}(x).$$

Let us write

$$\phi_{kn}(x) = \Phi_n^{2r}(x_{kn}, y)l_{kn}(x).$$

With these notations we define the operator

$$J_n^{(\alpha,\beta)}(f, x) = L(f, x) + \sum_{k=1}^{n}(f(x_{kn}) - L(f, x))\phi_{kn}(x), \qquad (5.9)$$

for $f \in C[-1,1]$.

Theorem 5.2.3 (Freud and Sharma, [126] and [127]). *Fix $\alpha, \beta > -1$ and r such that*

$$2r > \max\{4, \alpha + 5/2, \beta + 5/2\}.$$

For each n, let $J_n^{(\alpha,\beta)}$ be defined by (5.9). There exists a constant C such that, for each $f \in C[-1,1]$ and $x \in [-1,1]$,

$$|f(x) - J_n^{(\alpha,\beta)}(f, x)| \leq C\left(\omega\left(f, \frac{\sqrt{1-x^2}}{n}\right) + \omega\left(f, \frac{1}{n^2}\right)\right).$$

Proof. It is know that Jacobi polynomials satisfy the equation

$$P_n^{(\alpha,\beta)}(-x) = (-1)^n P_n^{(\beta,\alpha)}(x).$$

Thus

$$J_n^{(\beta,\alpha)}(f(-t),x) = J_n^{(\alpha,\beta)}(f,-x).$$

Set $g(x) = f(-x)$. Then for $x \in [-1,0)$, one has

$$| f(x) - J_n^{(\alpha,\beta)}(f,x) | = | f(x) - J_n^{(\beta,\alpha)}(f(-t),-x) | = | g(-x) - J_n^{(\beta,\alpha)}(g,-x) | .$$

Since $\omega(f,t) = \omega(g,t)$, we reduce the proof to the case $x \in [0,1]$ (of course, with different parameters, but it does not change the estimate).

In what follows we assume $x \in [0,1]$. Set

$$S_1(x) = \sum_{k=1}^{m} \phi_{kn}(x) \, | f(x) - f(x_{k,n}) |$$

$$S_2(x) = \frac{1+x}{2} \times | f(x) - f(1) | \times \left| 1 - \sum_{k=1}^{n} \phi_{kn}(x) \right|$$

and

$$S_3(x) = \frac{1-x}{2} \times | f(x) - f(-1) | \times \left| 1 - \sum_{k=1}^{n} \phi_{kn}(x) \right|.$$

Since

$$| f(x) - J_n^{(\alpha,\beta)}(f,x) | \le S_1(x) + S_2(x) + S_3(x),$$

we will estimate the last three terms.

Since $\Phi_m^{2r}(x,x) = \sum_{k=1}^{n} \phi_{nk}(x)$, it can be proved that

$$\frac{1}{2} < \Phi_m(x,x) \le 3 \quad \text{and} \quad \sqrt{1-x^2} \, |\Phi_m^2(x,x) - 1| \le \frac{4}{m}.$$

From these inequalities we obtain

$$S_2(x) \le \frac{1+x}{2} \omega\left(f, | x-1 | \right) \left| 1 - \Phi_m^{2r}(x,x) \right|$$

$$\le \frac{1+x}{2} \left(1 + n\frac{| x-1 |}{\sqrt{1-x^2}}\right) \omega\left(f, \frac{\sqrt{1-x^2}}{n}\right) \left| 1 - \Phi_m^{2r}(x,x) \right|$$

$$\le C\omega\left(f, \frac{\sqrt{1-x^2}}{n}\right).$$

For $S_3(x)$ we can use similar arguments.

The estimate for $S_1(x)$ is more complicated. Set $A = \{k : 0 \le \theta_{kn} \le 3\pi/4\}$ and $B = \{k : 3\pi/4 < \theta_{kn} \le \pi\}$ and split S_1 in two sums, $S_1 = \sum_{k \in A} + \sum_{k \in B}$.

We present the estimate for the second sum. For the first one some changes are needed.

If $\theta \in [0, \pi/2]$ and $\theta_{kn} \in (3\pi/4, \pi)$ $(x \in \cos\theta)$, then $\mid x - x_{n,k} \mid > c$. We need some properties of the zeros of Jacobi polynomials. If c is a positive constant, there exists a constant C_1 such that, for $cn^{-1} \leq \theta \leq \pi/2$

$$\mid P_n^{(\alpha,\beta)}(\cos\theta) \mid \leq C_1 \frac{1}{\sqrt{n}\,(\sin\theta)^{\alpha+1/2}}.$$

Moreover,

$$(P_n^{(\alpha,\beta)})'(\cos\theta_{kn}) \sim \frac{\sqrt{n}}{(\sin(\theta_{kn}))^{\beta+3/2}}, \qquad (\pi/2 \leq \theta_{kn} \leq \pi).$$

Since

$$\mid \Phi_m(x_{kn,x}) \mid \leq \frac{2}{m}\,\frac{\sin(\theta/2) + \sin(\theta_{kn}/2)}{\mid \cos\theta - \cos\theta_{kn} \mid},$$

there exists a constant C_2 such that

$$\mid \phi_{kn}(x) \mid = \mid l_{kn}(x) \mid \times \mid \Phi_m^{2r}(x_{kn}, x) \mid = \frac{\mid P_n^{(\alpha,\beta)}(x) \mid\mid \Phi_m^{2r}(x_{kn}, x) \mid}{\mid x - x_{kn} \mid\mid (P_n^{(\alpha,\beta)})'(x_{kn}) \mid}$$

$$\leq C_2 \frac{1}{n^{2r}}\,\frac{n^{\max\{\alpha,-1/2\}}}{\sqrt{n}}.$$

Finally, since $\mid f(x) - f(x_{kn}) \mid \leq \omega(f,2) \leq (1 + 2n^2)\omega(f, 1/n^2)$, we obtain

$$\sum_{k \in B} \leq C_3\,n^2\omega(f,n^{-2})\sum_{k \in B} n^{-2r-1/2+\max\{\alpha,-1/2\}}$$

$$\leq C_3\,n^2\omega(f,n^{-2})\,n^{-2r+3/2+\max\{\alpha,-1/2\}} \leq C_4\,\omega(f,n^{-2}),$$

if $-2r + 3/2 + \max\{\alpha, -1/2\} < 0$. $\qquad\qquad\qquad\qquad\qquad\qquad\qquad\qquad \square$

If, for $c > 0$ fixed, we chose ρ such that

$$n + 2rm - 1 < n(1 + 2r\rho) \leq n(1 + c),$$

then

$$J_n^{(\alpha,\beta)}(C[-1,1]) \subset \Pi_{n+2rm-1}.$$

The polynomials $J_n^{(\alpha,\beta)}(f)$ do not interpolate. But we can define

$$A_n^{(\alpha,\beta)}(f, x) = L(f, x) + \sum_{k=1}^{n}(f(y_{kn}) - L(f,x))\frac{\phi_{kn}(x)}{\Phi_m^{2r}(y_{kn}, y_{kn})}.$$

The new operators interpolate and an estimate like the one in the last theorem holds.

In 1983, Misra generalized Freud-Sharma operators $J_n^{(\alpha,\beta)}$ and $A_n^{(\alpha,\beta)}$ respectively without affecting their degree. The generalized operator is non-interpolatory while it produces Timan's estimate for $f^{(p)} \in C[-1,1]$.

Let $\{x_{k,n}\}$ be the zeros of the Jacobi polynomial $P_n^{(\alpha,\beta)}$, $\alpha, \beta > -1$, and denote by $l_{k,n}$ the fundamental polynomial of Lagrange interpolation based on these nodes. We shall denote $x_{k,n}$ by x_k, $l_{k,n}$ by l_k for the sake of convenience. Let $m = [n\rho]$, for some ρ, $0 < \rho < (r+p)^{-1}2^{-(p+1)}$, $p \geq 0$ where $2r > \max(4, \alpha + 5/2, \beta + 5/2)$. We set

$$\varphi_m(x,y) = \frac{1}{m} \frac{T_{m+1}(x)T_m(y) - T_{m+1}(y)T_m(x)}{x - y}$$

where $T_m(x) = \cos m\theta$, $x = \cos\theta$ so that ([5], p. 238)

$$\varphi_m(x,x) = \frac{1}{m}\left(m + \frac{1}{2} + \frac{1}{2}\frac{\sin(2m+1)\theta}{\sin\theta}\right).$$

Now, we introduce the polynomials $\psi_p(x,y)$ of degree $\leq 2^p m$ defined as follows:

$$\Psi_p(x,y) = \begin{cases} \varphi_m(x,y), & \text{if } p = 0, \\ \Psi_{p-1}(x,y)\overline{\Psi}_{p-1}(x,y), & \text{if } p \geq 1, \end{cases}$$

where $\overline{\Psi}_{p-1}(x,y) = 2 - \Psi_{p-1}(x,y)$.

Let

$$\lambda(f,x) = \frac{1}{2^{2p+1}} \sum_{i=1}^{p} \binom{2p+1}{i} \left((1+x)^{2p+1-i}(1-x)^i \sum_{j=0}^{p} \frac{(x-1)^j}{j!}f^{(j)}(1) \right.$$

$$\left. + (1-x)^{2p+1-i}(1+x)^i \sum_{j=0}^{p} \frac{(1+x)^j}{j!}f^{(j)}(-1) \right).$$

Now define

$$J_{n,p}^{(\alpha,\beta)}(f,x) = \lambda_p(f,x) + \sum_{k=1}^{n} \left(\sum_{i=0}^{p} \frac{(x-x_k)^i}{i!}f^{(i)}(x_k) - \lambda_p(f,x) \right) \Psi_p^{2r+2p}(x_k,x)l_k(x).$$

The operator $J_{n,p}^{(\alpha,\beta)}(f,x)$ is non-interpolatory and of degree $n + 3p + m(r + p)2^{p+1} \leq n(1+c)$, $c > 0$ being fixed.

Theorem 5.2.4 (Misra, [252]). *For $f \in C^p[-1,1]$,*

$$| f(x) - J_{n,p}^{(\alpha,\beta)}(f,x) | \leq C_p \left(\frac{\sqrt{1-x^2}}{n} + \frac{|x|}{n^2} \right)^p \omega(f^{(p)}, \Delta_n(x)).$$

Misra included a modification to obtain interpolatory polynomials. In the case $\alpha = \beta = -1/2$ he also proved a result on simultaneous approximation.

5.3 Gopengauz estimates

Lorentz and Steckin asked if it is possible to replace the inequality

$$| f(x) - p_n(x) | \le C_r \, \omega_r \left(f, \frac{\sqrt{1 - x^2}}{n} + \frac{1}{n^2} \right)$$

by

$$| f(x) - p_n(x) | \le C_r \, \omega_r \left(f, \frac{\sqrt{1 - x^2}}{n} \right).$$

This was shown to be possible in the case $r = 1$ by Teliakovskii [371]. In this section we present several different constructions which provide an inequality like the second one.

Two different constructions appeared in 1973, due to Saxena and Rodina. Saxena used a simple modification of the operators defined in (5.6).

Theorem 5.3.1 (Saxena, [326]). *For each n, let J_n be defined by (5.6) and set*

$$S_n(f, x) = J_n(f, x) + L(f - J_n(f), x),$$

where L is defined by (5.1). There exists a constant C such that, for $f \in C[-1, 1]$ and $x \in [-1, 1]$, one has

$$| f(x) - S_n(f, x) | \le C \omega \left(f, \frac{\sqrt{1 - x^2}}{n} \right).$$

Rodina used the Chebyshev polynomials U_n of second kind (5.5). Let $y_{k,n}$ be the roots of U_n. In this case the fundamental polynomials of Lagrange and the Hermite-Fejér formula can be written as

$$l_{k,n}(x) = \frac{(-1)^{k+1}(1 - y_{k,n}^2)U_n(x)}{(n+1)(x - y_{k,n})}, \qquad v_{k,n}(x) = 1 - \frac{3y_{k,n}(x - y_{k,n})}{1 - y_{k,n}^2}.$$

Now we set

$$\varphi_{k,n}(x) = \frac{1 - x^2}{1 - y_{k,n}^2} \left[l_{k,n}^4(x)v_{k,n}(x) + 2(x - y_{k,n})l_{k,n}^3(x)(1 - y_{k,n}^2)\psi_n(x, y_{k,n}) \right]$$

where

$$\psi_n(x, u) = \frac{2}{n+1} \sum_{r=1}^{n-1} U_r'(x)U_r(u).$$

Theorem 5.3.2 (Rodina, [313]). *For $n \in \mathbb{N}$ and $f \in C[-1, 1]$ define*

$$\Lambda_n(f, x) = L(f, x) + \sum_{k=1}^{n} [f(x_k) - L(f, x)]] \, \varphi_{k,n}(x).$$

There exists a constant C such that, for $n \in \mathbb{N}$, $f \in C[-1,1]$ and $x \in [-1,1]$ one has

$$| f(x) - \Lambda_n(f,x) | \leq C w \left(f, \frac{\sqrt{1-x^2}}{n+1} \right).$$

In the theorems given above the estimates are in terms of the first modulus of continuity. In 1975, DeVore was able to construct a sequence for which the estimate is given in terms of the second-order modulus.

Fix a sequence $\{K_n\}$ of non-negative trigonometric polynomials (deg $K_n \leq n$), such that

$$\int_{-\pi}^{\pi} | t |^j K_n(t) dt \leq C n^{-j}, \qquad 1 \leq j \leq 4.$$

For $h \in C_{2\pi}$ define

$$L_n(h,s) = \int_{-\pi}^{\pi} \left[-\Delta_t^4(h,s) + h(s) \right] K_n(t) \, dt. \tag{5.10}$$

For $f \in C[-1,1]$, let $P(f)$ be the polynomial of degree 1 which interpolates f at the points 1 and -1.

Define a sequence of linear operators by

$$\Lambda_n(f,x) = L_n \left(f(\cos(s)) - P(f, \cos s), \cos^{-1}(x) \right) + P(f,x).$$

Finally, define

$$M_n(f,x) = \Lambda_n(f,x) + U_n(f,x), \tag{5.11}$$

where U_n is the first degree polynomial which interpolates $f - \Lambda_n(f)$ at the points -1 and 1.

Theorem 5.3.3 (DeVore, [87] (see also DeVore [88]). *For each $n \geq 2$, let M_n be defined by (5.11). Then $M_n : C[-1,1] \to \mathbb{P}_n$ and, for every $f \in C[-1,1]$, one has*

$$| f(x) - M_n(f,x) | \leq C \omega_2 \left(f, \frac{\sqrt{1-x^2}}{n} \right), \qquad -1 \leq x \leq 1,$$

where the constant C does not depend on f or n.

Proof. Since L_n was constructed by convolution with a trigonometric kernel, ones has $M_n(f) \in \mathbb{P}_n$, for any $f \in C[-1,1]$. Taking into account the definition of U_n, we know that $M_n(f, \pm 1) = f(\pm 1)$.

First, we will verify that, for any $g \in W_{2,\infty}[-1,1]$,

$$| g(x) - M_n(g,x) | \leq C \|g''\|_\infty \frac{1-x^2}{n^2}, \qquad -1 \leq x \leq 1, \tag{5.12}$$

Fix $x \in [-1, 1]$. For an arbitrary function $g \in W_{2,\infty}[-1, 1]$, set $g_1 = g - P(g)$. Notice that $g_1^{(2)} = g^{(2)}$ and g_1' has a zero in $(-1, 1)$ (because $g_1(-1) = g_1(1) = 0$). Thus, by the mean value theorem we obtain

$$\|g_1\| = \sup_{x \in [-1,1]} |g_1(x) - g_1(-1)| \leq \sup_{x \in [-1,1]} (x+1)\|g_1'\| = 2\|g_1'\| \leq 4\|g_1^{(2)}\|.$$

Set $x = \cos s$ and $h(s) = g_1(\cos(s))$. Taking into account that $h(0) = 0$ and $h'(0) = 0$, we obtain the representation

$$h(s) = \int_0^s h'(t)dt = \int_0^s [h'(t) - h'(0)]dt$$

$$= \int_0^s \int_0^t h''(u)dudt = \int_0^s \int_0^t [\sin^2 u g_1^{(2)}(\cos u) - \cos u g_1'(\cos u)]dudt$$

$$= \int_0^s \int_0^t [\sin^2 u g_1^{(2)}(\cos u)]dudt$$

$$- \int_0^s \int_0^t [\cos u g_1'(\cos u) - \cos(\pi/2)g_1'(\cos(\pi/2))]dudt$$

$$= \int_0^s \int_0^t [\sin^2 u g_1^{(2)}(\cos u)]dudt$$

$$+ \int_0^s \int_0^t \int_{\pi/2}^u \sin v [\cos v g_1^{(2)}(\cos v) + g_1'(\cos v)]dvdudt.$$

Set

$$H_1(s) = \int_0^s \int_0^t [\sin^2 u g_1^{(2)}(\cos u)]dudt$$

and

$$H_2(s) = \int_0^s \int_0^t \int_{\pi/2}^u \sin v [\cos v g_1^{(2)}(\cos v) + g_1'(\cos v)]dvdudt.$$

We need some properties of the differences of H_1 and H_2. Notice that, since $|H_1^{(2)}(s)| \leq \|g_1^{(2)}\| \sin^2 s = \|g^{(2)}\| \sin^2 s$, one has

$$|\Delta_t^4 H_1(s)| \leq 4|\Delta_t^2 H_1(s)| = 4\left|\int_0^t \int_0^t H_1^{(2)}(s + t_1 + t_2)dt_1 dt_2\right|$$

$$\leq 4\|g^{(2)}\|t^2 \sup_{|u| \leq 2|t|} \sin^2(s + u) \leq 48\|g^{(2)}\|t^2(t^2 + \sin^2 s),$$

because, taking into account that

$$|\sin(s + u)| \leq |\sin s| + |\sin u| \leq |\sin s| + |u|,$$

if $|u| \leq 2|t|$, then

$$\sin^2(s + u) \leq \sin^2 s + 2|u \sin s| + u^2 \leq 3(\sin^2 s + u^2) \leq 12(\sin^2 s + t^2).$$

On the other hand, since

$$| H_2^{(3)}(s) | \leq (\|g_1^{(2)}\| + \|g_1'\|) \mid \sin s \mid \leq 3\|g^{(2)}\| \mid \sin s \mid,$$

one has

$$| \Delta_t^4 H_2(s) | \leq 2 | \Delta_t^3 H_2(s) | = 2 \left| \int_0^t \int_0^t \int_0^t H_2^{(3)}(s + t_1 + t_2 + t_3) dt_1 dt_2 dt_3 \right|$$

$$\leq C_1 \|g^{(2)}\| \mid t \mid^3 \sup_{|u| \leq 3|t|} \mid \sin(s + u) \mid \leq C_2 \|g^{(2)}\| \mid t \mid^3 (\mid t \mid + \mid \sin s \mid).$$

With the estimates given above we obtain $(x = \cos s)$

$$| g(x) - \Lambda_n(g, x) | = | g_1(x) - L_n(g_1, \cos^{-1} x) | = | h(s) - L_n(h, s) |$$

$$\leq | H_1(s) - L_n(H_1, s) | + | H_2(s) - L_n(H_2, s) |$$

$$\leq \int_{-\pi}^{\pi} | \Delta_t^4 H_1(s) | K_n(t) \, dt + \int_{-\pi}^{\pi} | \Delta_t^4 H_2(s) | K_n(t) \, dt$$

$$\leq C_3 \|g^{(2)}\| \int_{-\pi}^{\pi} (t^2(t^2 + \sin^2 s) + \mid t \mid^3 (\mid t \mid + \mid \sin s \mid)) K_n(t) \, dt$$

$$\leq C_4 \|g^{(2)}\| \left(\frac{1}{n^4} + \frac{\sin^2 s}{n^2} + \frac{\mid \sin s \mid}{n^3} \right)$$

$$\leq C_5 \|g^{(2)}\| \left(\frac{1}{n^2} + \frac{\mid \sin s \mid}{n} \right)^2.$$

In particular $|g(\pm 1) - \Lambda_n(g, \pm 1)| \leq C \|g^{(2)}\| n^{-4}$. Therefore

$$\|U_n(g)\| \leq C_6 \|g^{(2)}\| n^{-4}.$$

Moreover, since $U_n(g)$ is a first degree polynomial

$$\|U_n'(g)\| \leq C_7 \|g^{(2)}\| n^{-4}.$$

The last inequalities provide the estimate

$$| g(x) - M_n(g, x) | \leq C_8 \|g^{(2)}\| \left(\frac{1}{n^2} + \frac{\mid \sin s \mid}{n} \right)^2.$$

If $n^{-1} \leq \mid \sin s \mid$, then from the last inequality we obtain

$$| g(x) - M_n(g, x) | \leq C_9 \|g^{(2)}\| \left(\frac{\mid \sin s \mid}{n} \right)^2 = C_9 \|g^{(2)}\| \left(\frac{\sqrt{1 - x^2}}{n} \right)^2.$$

Now we will verify the inequality when $\mid \sin s \mid < n^{-1}$. We consider the case $0 \leq x < 1$ (the result for the other case follows analogously).

Taking into account that $g(1) - M_n(g, 1) = 0$, there exists $y \in (x, 1)$ such that

$$
\begin{aligned}
\mid g(x) - M_n(g, x) \mid &= \mid g(x) - M_n(g, x) - g(1) + M_n(1) \mid \\
&= (1 - x) \mid g'(y) - M_n'(g, y) \mid \\
&\leq (1 - x) \mid g'(y) - \Lambda_n'(g, y) \mid + (1 - x)\|U_n'(g)\| \\
&\leq (1 - x^2) \mid g_1'(y) - L_n'(g_1, y) \mid + C_{10}\|g^{(2)}\|\frac{1 - x^2}{n^2}.
\end{aligned}
$$

Thus, in order to finish this part of the proof, we need to estimate $g_1'(y) - \Lambda_n'(g_1, y)$ when $\mid \sin s \mid < 1/n$ $(x = \cos s)$ and $x < y < 1$.

Set $y = \cos u$ (notice that $0 < u \leq \pi/2$). Since $h - L_n(h)$ is an even function, $h'(0) - L_n'(h, 0) = 0$. Therefore, by the mean value theorem, there exists $v \in (0, u)$ such that

$$
\begin{aligned}
\mid g_1'(y) - L_n'(g_1, y) \mid &= \frac{1}{\sin u} \mid h'(u) - L_n(h', u) \mid \\
&= \frac{1}{\sin u} \mid h^{(2)}(v) - L_n(h^{(2)}, v) \mid v \leq C_{11} \mid h^{(2)}(v) - L_n(h^{(2)}, v) \mid \\
&\leq C_{11} \left(\mid H_1^{(2)}(v) - L_n(H_1^{(2)}, v) \mid + \mid H_2^{(2)}(v) - L_n(H_2^{(2)}, v) \mid \right) \\
&\leq C_{11} \left(\int_{-\pi}^{\pi} \mid \Delta_t^4 H_1^{(2)}(v) \mid K_n(t)\, dt + \int_{-\pi}^{\pi} \mid \Delta_t^4 H_2^{(2)}(v) \mid K_n(t)\, dt \right) \\
&\leq C_{12}\|g^{(2)}\|\frac{1}{n^2}
\end{aligned}
$$

because

$$
\begin{aligned}
\mid \Delta_t^4 H_1^{(2)}(v) \mid &\leq C_{13}\|g^{(2)}\| \sup_{|w| \leq 4|t|} \sin^2(v + w) \\
&\leq C_{14}\|g^{(2)}\|(t^2 + \sin^2 v) \leq C_{14}\|g^{(2)}\|(t^2 + \sin^2 u) \\
&\leq C_{14}\|g^{(2)}\|(t^2 + n^{-2})
\end{aligned}
$$

and

$$
\begin{aligned}
\mid \Delta_t^4 H_2^{(2)}(v) \mid &\leq C_{15} \mid \Delta_t H_2^{(2)}(v) \mid = C_{15} \left| \int_0^t H_2^{(3)}(v + t_1)dt_1 \right| \\
&\leq C_{15}\|g^{(2)}\| \mid t \mid \sup_{|w| \leq |t|} \mid \sin(v + w) \mid \leq C_{16}\|g^{(2)}\| \mid t \mid (\mid t \mid + n^{-1}).
\end{aligned}
$$

We have proved (5.12).

Now, in order to obtain the general result we use standard arguments. In particular, we only need an estimate in terms of the K-functional. Taking into account that the sequence of operators $\{M_n\}$ is uniformly bounded, if $f \in C[-1, 1]$,

$x \in [-1, 1]$ and $g \in W_{2,\infty}[-1, 1]$ one has

$$| f(x) - M_n(f, x) | \le C_{17} (\|f - g\| + | g(x) - M_n(g, x) |)$$

$$\le C_{18} \left(\|f - g\| + \frac{1 - x^2}{n^2} \|g^{(2)}\| \right).$$

Therefore

$$| f(x) - M_n(f, x) | \le C_{17} K \left(f, \frac{1 - x^2}{n^2}, C[-1, 1], W_{2,\infty}[-1, 1] \right). \qquad \square$$

In 1979 Varma and Mills [395] also gave an interpolatory proof for a Telia-kovskii-type estimate, but for the first modulus. The use of the interpolation process of Bernstein will be analyzed in the next section.

Later, in 1986, Kis and Szabados [198] constructed a family of operators which depends on several parameters (j, k, l and m). The estimates should be viewed as $m \to \infty$ (or $n \to \infty$), while the other parameters (j, k, l) remain fixed. For a certain choice of the parameters the operator converges in the order of best approximation. They obtained the Jackson, Timan, and Teliakowskii-Gopengauz theorems with explicit constants.

Fix $j, k, l, m \in \mathbb{N}$ such that

$$n = \frac{1}{2}(jm + km - k + l - 1) \in \mathbb{N}_0$$

and define

$$t_v = \frac{2\pi v}{jm}, \qquad (v \in \mathbb{Z}),$$

$$s_{j,k,l,m}(t) = \frac{\sin(jmt/2) \sin^k(mt/2) \cos^l(t/2)}{jm^{k+1} \sin^{k+1}(t/2)}, \qquad \sin(t/2) \ne 0,$$

$$s_{j,k,l,m}(t) = \lim_{\tau \to t} s_{j,k,l,m}(\tau) = 1, \qquad \sin(t/2) = 0$$

and

$$S_{j,k,l,m}(g, t) = \sum_{v=}^{jm-1} g(t_v) s_{j,k,l,m}(t - t_v).$$

Notice that $S_{1,0,0,m}(t)$ (m odd) is the Dirichlet kernel, $S_{1,1,0,m}(t)$ is the Fejér kernel, $S_{3,1,0,m}(t)$ is the de la Vallée-Poussin kernel, and $S_{1,3,0,m}(t)$ is the Jackson kernel. Operators $S_{3,1,0,m}$ were previously studied by Szabados in [361].

Let us also define

$$L_{j,k,l,m}(t) = \sum_{v=0}^{jm-1} | s_{j,k,l,m}(t - t_v) |,$$

$$M_{j,k,l,m}(t) = \sum_{v=0}^{jm-1} \left| \frac{\sin(m(t - t_v)/2)}{m \sin((t - t_v)/2)} \right|^k |\cos((t - t_v)/2)|$$

and

$$q = \frac{1}{2}(jm - km + k - l - 1).$$

Now, for $F : [-1,1] \to \mathbb{R}$ and $x = \cos t$ define

$$P_{j,k,lm}(f,x) = S_{j,k,lm}(f \circ \cos, t)$$

and set

$$x_v = \cos \frac{2\pi v}{jm}, \qquad (v \in \mathbb{Z}).$$

Theorem 5.3.4 (Kis and Szabados, [198]). *Assume that $q \geq 0$ and fix $f \in C[-1,1]$.*

(i) $\qquad\qquad P_{j,k,lm}(f) \in \mathbb{P}_n \quad$ and $\quad P_{j,k,lm}(f,x_v) = f(x_v), \ v \in \mathbb{Z}.$

(ii) $\qquad\qquad \|f - P_{j,\,k,\,l,\,m}(f)\| \leq (1 + L_{j,\,k,\,l,\,m}(t))\, E_q(f).$

(iii) *If $k \geq 1$, then*

$$\|f - P_{j,\,k,\,l,\,m}(f)\| \leq \left(L_{j,k,l,m}(t) + \frac{\pi}{2} M_{j,k,l,m}(t) \right) \omega \left(f, \frac{\pi}{jm} \right).$$

(iv) *If $k \geq 2$, $q \geq 0$ and $x \in [-1,1]$, then*

$$|f(x) - P_{j,\,k,\,l,\,m}(f,x)| \leq \left(L_{j,\,k,\,l,\,m}(t) + \frac{\pi M_{j,\,k,\,l+1,\,m}(t)}{2} \right) \omega \left(f, \frac{\pi\sqrt{1-x^2}}{jm} \right)$$

$$+ \left(L_{j,\,k,\,l,\,m}(t) + \frac{2j}{\pi^2} M_{j,\,k-1,\,l,m}(t) \right) \omega \left(f, \frac{\pi\,|x|}{j^2 m^2} \right).$$

(v) *If j is odd, $k \geq 2$ is even, $q \geq 0$ and $x \in [-1,1]$, then*

$$|f(x) - P_{j,\,k,\,l,\,m}(f,x)| \leq L_{j,\,k,\,l,\,m}(t)\omega \left(f, \frac{\pi\sqrt{1-x^2}}{m} \right)$$

$$+ \left(2L_{j,\,k,\,l,\,m}(t) + \frac{2}{\pi j} M_{j,\,k,\,l+1,\,m}(t) \right) \omega \left(f, \frac{2\pi^2\,|x|}{m^2} \right).$$

(vi) *If j or m is even, $k \geq 2$, $q \geq 0$ and $x \in [-1,1]$, then*

$$|f(x) - P_{j,\,k,\,l,\,m}(f,x)| \leq \left(L_{j,\,k,\,l,\,m}(t) + \frac{2}{\pi j} M_{j,\,k,\,l+1,\,m}(t) \right.$$

$$\left. + \frac{j\,|x|}{\pi} M_{j,\,k-1,\,l,\,m}(t) \right) \omega \left(f, \frac{\pi\sqrt{1-x^2}}{jm} \right).$$

(vii) *If j odd, m and $k \geq 2$ are even, $q \geq 0$ and $x \in [-1,1]$, then*

$$| f(x) - P_{j,k,l,m}(f,x) | \leq \left(2L_{j,k,l,m}(t) + \left(1 + \frac{|x|}{2} \right) M_{j,k,l,m}(t) \right)$$
$$\times \omega \left(f, \frac{2\pi\sqrt{1-x^2}}{m} \right).$$

As corollaries one has

$$\|f - P_{2,2,1,m}(f)\| \leq \left(\frac{2}{\sqrt{3}} + \frac{4}{\pi} \right) \omega \left(f, \frac{\pi}{n+1} \right),$$

and

$$\|f - P_{3,3,2,m}(f)\| \leq \left(\frac{11}{9} + \frac{2\sqrt{6}+9}{\pi} \right) \omega \left(f, \frac{\pi\sqrt{1-x^2}}{n+1} \right).$$

Theorem 5.3.5 (Kis and Szabados, [198]). *Given $0 < \varepsilon \leq 1$, for each $n \geq 20/\varepsilon^2$ and $f \in C[-1,1]$, there exists a polynomial $P_n \in \mathbb{P}_{n(1+\varepsilon)}$ such that $P_n(f)$ interpolates f in at least n points and*

$$|f(x) - P_n(f,x)| \leq \frac{13}{\varepsilon^2} \omega_1 \left(f, \frac{\pi\sqrt{1-x^2}}{2n} \right).$$

This answers a question about the construction of linear operators of minimal degree compared to the number of points of interpolation, that at the same time realize the Teliakovskii-Gopengauz estimate.

For classical Hermite interpolation, Gopengauz found a point-wise estimate of the remainder of an interpolation formula using two multiple nodes at ± 1.

Theorem 5.3.6 (Gopengauz, [152]). *Fix $r \in \mathbb{N}$, $f \in C^{r-1}[-1,1]$ and let $P \in \Pi_{2r-1}$ be a Hermite interpolation polynomial given by $f^{(v)}(-1) = p^{(v)}(-1)$ and $f^{(v)}(1) = p^{(v)}(1)$ for $v = 0,1,2,\ldots,r-1$. Then for all $x \in [-1,1]$,*

$$| f(x) - p(x) | \leq C_r (1-x^2)^{r-1} \omega_{r+1} \left(f^{(r-1)}, \frac{2}{r+1}(1-x^2)^{1/(r+1)} \right)$$

where the constant C_r depends only on r.

5.4 Bernstein interpolation process

We know that for the Lagrange interpolation L_n the Lebesgue constant is not bounded with respect to n. To avoid this drawback some modifications are considered. One of them was suggested by Bernstein in 1930, who asked whether it is possible, for a given $\lambda \in (1,2)$, to construct a Lagrange-like interpolation polynomial Q_n of degree $\leq \lambda N$, such that for every $f \in C[-1,1]$ the polynomial Q_n interpolates at least at N points of f and $\|Q_n - f\| \to 0$ as $N \to \infty$.

In fact we have several problems: 1) Is such a construction possible? 2) If a construction is possible, give a clear description. 3) Find a good estimate for the rate of convergence. In 1) we do not ask for a clear construction.

An answer to the first problem was given by Erdös in 1943 (see [114]). He provided a characterization.

Suppose all nodes $x_{k,n}$ lie in $(-1,1)$ and $x_{k,n} = \cos\theta_{k,n}$. Denote by $N(a_n,b_n)$ the numbers of points $\theta_{k,n}$ in the interval (a_n,b_n), where $0 \leq a_n < b_n \leq \pi$. If $n(b_n - a_n) \to 0$, then the Erdös conditions are

$$\lim_{n\to\infty} \sup \frac{N(a_n,b_n)}{n(b_n - a_n)} \leq \frac{1}{\pi}$$

and, for each i,

$$\lim_{n\to\infty} \inf n(\theta_{i,n} - \theta_{i+1,n}) > 0.$$

Some further investigations can be found in a paper of Vértesi [400] who, among others, proved that if a system of nodes $\{x_{k,n}\}$ satisfies the so-called Erdös condition, then there exists a linear operator L_n such that $L_n(f,x)$ is an algebraic polynomial of degree $(1+c)n$ for every $f \in C^r[-1,1]$ and some $c > 0$.

Definition 5.4.1. Fix $c > 0$. A system of points $\{x_{k,n}\}$ ($n \in \mathbb{N}$, $1 \leq k \leq n$) is called of *Bernstein-Erdös type*, if there exists a sequence of linear operators $\{L_n\}$ ($L_n : C[-1,1] \to C[-1,1]$) for which the following three conditions hold:

$$L_n(C[-1,1]) \subset \Pi_{n(1+c)}, \tag{5.13}$$
$$L_n(f, x_{k,n}) = f(x_{k,n}), \quad \text{for all} \quad f \in C[-1,1] \quad \text{and} \quad 1 \leq k \leq n, \tag{5.14}$$
$$\lim_{n\to\infty} L_n(f,x) = f(x). \tag{5.15}$$

The system is called *well approximating* if there is a sequence $\{L_n\}$ which satisfies (i) and (ii) and

$$\|f - L_n\| \leq C E_n(f),$$

where C depends on c and $\{x_{k,n}\}$, but not on f.

In 1932, Bernstein provided a constructive solution to the problem. He proved that the zeros of Chebyshev polynomials $\{x_{k,n}\}$ is a system of Bernstein-Erdös type. Fix $\lambda > 1$ and set $l = \lambda/(2(\lambda - 1))$. Notice that

$$\frac{2l}{2l - 1} = \lambda.$$

Distribute the nodes $x_{1,n} > x_{2,n} > \cdots > x_{n,n}$ in groups of $2l$ neighbor nodes. If n is not divisible by $2l$, the extreme groups may have less than $2l$ elements. To each one of the obtained $2l - 1$ groups we associate a value $f(x_{k,n}) = A_k$. These correspond to the first $2l - 1$ groups. If $k = 2ls$, the value A_{2ls} is defined by

$$A_{2(s-1)+l+1} + A_{2(s-1)+l+3} + \cdots + A_{2ls-1} = A_{2(s-1)+l+2} + A_{2(s-1)+l+4} + \cdots + A_{2ls}.$$

With this construction the interpolation formula is defined by

$$Q_n(f, x) = T_n(x) \sum_{k=1}^{n} \frac{A_k}{(x - x_{k,n})T_n'(x_{k,n})}. \tag{5.16}$$

We have that $Q_n(f)$ is a polynomial of degree not greater than $n - 1$ and if N is the number of points x where $Q_n(f, x) = f(x)$, then $N \geq n(2l - 1)/2l$.

Theorem 5.4.2 (Bernstein, [28]). *Assume that, for each n, Q_n is defined by (5.16). For each $f \in C[-1, 1]$, one has $\|f - Q_n(f)\| \to 0$.*

Bernstein also noticed that, if we know the values of the function f at the nodes $x_{k,n}$, we can obtain an interpolation formula with a better rate of convergence. We do not need to use all the given values of the function; we put the values in groups of $2l$ nodes and take one of the values of the function in each one of the obtained groups. The restriction we need is that the sum of the values with even index be equal to the sum of the values with odd index. In this construction nothing is said concerning which value of the function we will choose. For instance, one can take the mean value.

For $l = 1$, this idea leads to formulation of the operator

$$S_n(f, x) = \frac{T_n(x)}{2n} \sum_{k=1}^{n} (-1)^{k+1} \frac{(f(x_{k,n}) + f(x_{k+1,n}))\sqrt{1 - x_{k,n}^2}}{x - x_{k,n}}$$

where we consider $f(x_{n+1,n}) = f(x_{n,n})$.

Let us present another example. Set

$$\varphi_{1,n}(x) = \frac{3l_{1,n}(x) + l_{2,n}(x)}{4}, \qquad \varphi_{n-1,n}(x) = \frac{3l_{n,n}(x) + l_{n-1,n}(x)}{4} \tag{5.17}$$

and

$$\varphi_{k,n}(x) = \frac{l_{k-1,n}(x) + 2l_{k,n}(x) + l_{k+1,n}(x)}{4}, \qquad k = 2, \ldots, n - 1. \tag{5.18}$$

Then we define

$$R_n(f, x) = \sum_{k=1}^{n} f(x_k)\varphi_{k,n}(x). \tag{5.19}$$

Theorem 5.4.3 (Freud, [125]). *Given $c > 0$, for each triangular matrix satisfying Erdös's conditions one can find a sequence $\{A_n\}$ satisfying Bernstein's conditions (5.13) and (5.14) such that*

$$| f(x) - A_n(f, x) | \leq K(c)E_{n-1}(f).$$

Theorem 5.4.4 ([125]). *A sequence $\{x_{kn}\}$ is well approximating if and only if it is of Bernstein-Erdös type.*

5.4.1   Bernstein first interpolation operators

We will refer to (5.19) as the Bernstein first interpolation operators.

In 1973, Kis gave an estimate of the constant for one of the Bernstein operators.

Theorem 5.4.5 (Kis, [197]). *Assume that, for each n, R_n is defined by (5.19). For each $f \in [-1, 1]$, one has*

$$| f(x) - R_n(f, x) | \leq \frac{13}{3\pi} w \left(f, \frac{2\pi}{2n + 1} \right).$$

In 1976, Varma improved Theorem 5.4.2 by considering the rate of convergence.

Theorem 5.4.6 (Varma, [393]). *Assume that, for each n, R_n is defined by (5.19). There exists a positive constant C, such that for each $f \in C[-1, 1]$,*

$$| f(x) - R_n(f, x) | \leq C \left(w \left(f, \frac{\sqrt{1 - x^2}}{n} \right) + w \left(f, \frac{1}{n^2} \right) \right). \tag{5.20}$$

In 1989, Jiaxing proved a theorem that gives an estimate of Bernstein operators (5.19) for differentiable functions.

Theorem 5.4.7 (Jiaxing, [180]). *Assume that for each n, R_n is defined by (5.19). There exists a constant C such that, if $f \in C^1[-1, 1]$ and $x \in [-1, 1]$, then*

$$| f(x) - F_n(f, x) | \leq C \left(\frac{1}{n} w \left(f', \frac{\sqrt{1 - x^2}}{n} + \frac{1}{n^2} \right) + \frac{\| f' \|}{n^2} \right).$$

5.4.2   Chebyshev polynomials of second type

In 1978, Varma proved a result similar to Theorem 5.4.6 when the zeros of Chebyshev polynomials are changed to zeros of Chebyshev polynomials of the second kind. Set

$$\theta_k = \frac{k\pi}{n + 1}, \quad t_k = \cos \theta_k, \quad k = 1, \ldots, n,$$

$$\mu_k(x) = \frac{(-1)^{k+1}(1 - t_k^2)}{n + 1} \frac{U_n(x)}{x - t_k}, \quad k = 1, \ldots, n,$$

$$m_1(x) = \frac{3\mu_1(x) + \mu_2(x)}{4}, \quad m_n(x) = \frac{\mu_{n-1}(x) + 3\mu_n(x)}{4},$$

$$m_k(x) = \frac{\mu_{k-1}(x) + 2\mu_k(x) + \mu_{k+1}(x)}{4}, \quad k = 2, \ldots, n-2,$$

$$P_1(x) = m_1(x) + \frac{1}{2}m_2(x), \quad P_{n-1}(x) = \frac{1}{2}m_{n-1}(x), \quad P_n(x) = m_n(x),$$

$$P_k(x) = \frac{1}{2}(m_k(x) + m_{k+1}(x)), \quad k = 2, \ldots, n-2.$$

Let us set

$$A_n(f, x) = \sum_{k=1}^{n} f(t_k) m_k(x) \tag{5.21}$$

and

$$B_n(f, x) = \sum_{k=1}^{n} f(t_k) P_k(x).$$

Theorem 5.4.8 (Varma, [394]). *Let the operators $\{A_n\}$ and $\{B_n\}$ be defined as above. There exist constants C_1 and C_2 such that, for $f \in C[-1, 1]$, $n \in \mathbb{N}$ and $x \in [-1, 1]$,*

$$|f(x) - A_n(f, x)| \leq C_1 \omega\left(f, \frac{1}{n}\right)$$

and

$$|f(x) - B_n(f, x)| \leq C_1 \left(\omega\left(f, \frac{\sqrt{1-x^2}}{n}\right) + \omega\left(f, \frac{1}{n^2}\right)\right).$$

For the operators A_n, Jiazing obtained an analogue to Theorem 5.4.7.

Theorem 5.4.9 (Jiaxing, [181]). *Assume that for each n, A_n is defined by (5.21). There exists a constant C such that, if $f \in C^1[-1, 1]$ and $x \in [-1, 1]$, then*

$$|f(x) - A_n(f, x)| \leq C\left(\frac{1}{n}\omega\left(f', \frac{\sqrt{1-x^2}}{n} + \frac{1}{n^2}\right) + \frac{\|f'\|}{n^2}\right).$$

In 1982, Chauhan presented Teliakovskii-Gopengauz's theorems (in terms of the first modulus of continuity) taking nodes of interpolation at the roots of U_n including points ± 1. Set

$$t_k = \frac{k\pi}{n+1}, \quad x_k = \cos t_k, \quad 0 \leq k \leq n+1,$$

$$l_k(x) = \frac{(-1)^{k+1}(1-x)^2 U_n(x)}{(n+1)(x-x_k)}, \quad 1 \leq k \leq n,$$

$$l_0(x) = \frac{1+x}{2}\frac{U_n(x)}{n+1},$$

$$l_{n+1}(x) = (-1)^n \frac{1+x}{2}\frac{U_n(x)}{n+1}.$$

Now, define the polynomials

$$V_n(f, x) = \sum_{k=0}^{n+1} f(x_k) v_k(x) \quad \text{and} \quad Q_n(f, x) = \sum_{k=0}^{n} f(x_k) q_k(x)$$

where

$$v_0(x) = l_0(x), \qquad v_{n+1}(x) = l_{n+1}(x),$$

$$v_1(x) = \frac{3l_1(x) + l_2(x)}{4}, \qquad v_n(x) = \frac{l_{n-1}(x) + l_n(x)}{4},$$

$$v_k(x) = \frac{l_{k-1}(x) + 2l_k(x) + l_{k+1}(x)}{4}, \qquad 2 \le k \le n - 1,$$

$$q_0(x) = l_0(x), \qquad q_{n+1}(x) = l_{n+1}(x),$$

$$q_1(x) = \frac{7l_1(x) + 4l_2(x) + l_3(x)}{8} = v_1(x) + \frac{1}{2} v_2(x),$$

$$q_k(x) = \frac{l_{k-1}(x) + 3l_k(x) + l_{k+1}(x) + l_{k+2}(x)}{8}$$

$$= \frac{1}{2}[v_k(x) + v_{k+1}(x)] \qquad 2 \le k \le n - 2,$$

$$q_{n-1}(x) = \frac{l_{n-2}(x) + 2l_{n-1}(x) + l_n(x)}{8} = \frac{1}{2} v_{n-1}(x),$$

$$q_n(x) = \frac{l_{n-1}(x) + 3l_n(x)}{4} = v_n(x).$$

Theorem 5.4.10 (Chauhan, [74]). *There exist constants C_1 and C_2 such that, for each n, $f \in C[-1, 1]$ and $x \in [-1, 1]$,*

$$| f(x) - V_n(f, x) | \le C_1 \, \omega \left(f, \frac{1}{n} \right)$$

and

$$| f(x) - Q_n(f, x) | \le C_2 \, \omega \left(f, \frac{\sqrt{1 - x^2}}{n} \right).$$

In [413] Xie and Zhou presented a modification of Lagrange interpolation based on the zeros of the Chebyshev polynomial of the second kind.

5.4.3 General Bernstein operators

In 1996, Jiaxing modified the Bernstein process in order to consider derivatives up to order 3. Let the functions $\{\varphi_{k,n}\}$ be given by (5.17) and (5.18). Set

$$\psi_{1,n}(x) = \frac{5\varphi_{1,n}(x) - \varphi_{2,n}(x)}{4}, \qquad \psi_{n-1,n}(x) = \frac{5\varphi_{n,n}(x) - \varphi_{n-1,n}(x)}{4} \qquad (5.22)$$

and

$$\psi_{k,n}(x) = \frac{-\varphi_{k-1,n}(x) + 6l_{k,n}(x) - \varphi_{k+1,n}(x)}{4}, \qquad k = 2, \ldots, n-1. \qquad (5.23)$$

Now consider the operators

$$H_n(f, x) = \sum_{k=1}^{n} f(x_{k,n})\psi_{k,n}(x). \qquad (5.24)$$

Theorem 5.4.11 (Jiaxing, [183]). *Let H_n be defined by (5.24). If $f \in C^j[-1,1]$ $(0 \le j \le 3)$, then there exists a constant $C(f)$ such that*

$$\|f - H_n(f)\| \le C(f) \left(\frac{1}{n^j} \omega \left(f^{(j)}, \frac{1}{n} \right) + \frac{1}{n^{j+1}} \right).$$

The highest convergence order for H_n is n^{-4}. Jiaxing obtained a better estimate using the Ditzian-Totik modulus and the results of Ditzian and Jiang in [99].

Theorem 5.4.12 ([183]). *Let the sequence $\{H_n\}$ be defined by (5.24) and fix $\in [0,1]$. There exists a constant C such that $f \in C[-1,1]$ and $x \in [-1,1]$, thus one has*

$$| f(x) - H_n(f, x) | \le C\omega_\varphi^\lambda \left(f, \frac{1}{n}(\delta_n(x))^{1-\lambda} \right).$$

According to Jiaxing and Jichang [182] in 1993, Zhu obtained an estimate for the general Bernstein construction with Chebyshev nodes.

Theorem 5.4.13 (Zhu, [419]). *Let R_n be given by (5.16). For $f \in C[-1,1]$ and $x \in [-1,1]$, one has*

$$| f(x) - R_n(x) | \le C \left\{ \omega \left(f, \frac{|x|}{2} |\theta - \theta_{k_0}|^2 + \sqrt{1-x^2} |\theta - \theta_{k_0}| \right) \right.$$

$$+ |T_n(x)| \, \omega \left(f, \frac{\sqrt{1-x^2}}{n} + \frac{1}{n^2} \right)$$

$$\left. + \frac{|T_n(x)|}{n} \int_{1/n}^1 \frac{\omega(f, |x| t^2 + \sqrt{1-x^2} t)}{t^2} dt \right\},$$

where $x_{k_0} = \cos\theta_{k_0}$ is the nearest node to $x = \cos\theta$, and $k_0/2l$ is not an integer; C is a constant which depends only on l.

In this case the higher order of convergence can not exceed $1/n$. Jiaxing and Jichang presented another construction with a better rate of convergence for continuous and derivable functions.

Consider again the Chebyshev nodes $x_{k,n}$. Divide $x_{n,n} < x_{n-1,n} < \cdots < x_{3,n} < x_{2,n}$ according to $2l$. We have $n = 2ls + 2 + r$, $s \in \mathbb{N}$ and $0 \le r < 2l$. At the $2lt + 1$th nodes, $t = 1, 2, \ldots, s$ the value of $H_n(f, x)$ is

$$B_{2lt+1} = f(x_{2lt+1} + \sum_{p=1}^{l}(f(x_{2l(t-1)+2p} - f(x_{2l(t-1)+2p+1})$$

$$+ \frac{1}{4}\sum_{p=1}^{2}(f_{2lt+p}) - f(x_{2l(t-1)+p}).$$

At other nodes, the value of $H_n(f, x)$ is equal to $f(x)$.

Now define

$$H_n(f, x) = \sum B_k \mu_k(x)$$

where B_k is given as above when $k = 2lt + 1$, $t = 1, 2, \ldots, s$ and $B_k = f(x_k)$ otherwise.

It is known that $H_n(f, x)$ is a polynomial of degree $M = n - 1$, and $H_n(f)$ and f coincide at $G > (2l - 1)n/2l$ nodes; $M/G < 2l/(2l - 1) = \lambda$.

Theorem 5.4.14 (Jiaxing and Jichang, [182]). *There exists a constant C such that, for $f \in C[-1, 1]$ and $x \in [-1, 1]$,*

$$| f(x) - H_n(f, x) | \le C\omega\left(f, \frac{1}{n}\right).$$

Moreover, if $f \in C^1[-1, 1]$, then

$$| f(x) - H_n(f, x) | \le C\left(\frac{1}{n}\omega\left(f, \frac{1}{n}\right) + \frac{\|f'\|}{n^2}\right).$$

The paper also contains a result for functions $f \in C^2[-1, 1]$. The authors remarked that the order of convergence can not exceed $1/n^2$.

In the last section we recall that Varman and Mills proved that, with a Bernstein-type interpolation operator, we can obtain a Teliakovskii-type estimate [395].

5.4.4   Other modifications

In 1983, Chauhan observed that interpolating at one end of the interval did not considerably improve the estimate. Thus he proposed another construction. Set $t_{kn} = (2k - 1)\pi/(2n)$, $1 \le k \le n$ and $x_{kn} = \cos t_{kn}$. For $1 \le k \le 2n$, let

$$l_k(t) = \frac{\sin n(t - t_k)\cos((t - t_{kn})/2)}{2nn \sin(t - t_{kn})/2}$$

$$= \frac{1}{2n}\left(1 + \sum_{j=1}^{n-1}\cos j(t - t_{kn}) + \cos n(t - t_{kn})\right)$$

and
$$s_k(t) = 4t_k^3(t) - 3t_k^r(t).$$

Now, for $f : [-1,1] \to \mathbb{R}$, define

$$R_n(f,x) = L(f,x) + \sum_{k=1}^{n} (f(x_{kn}) + L(f,x)) \, r_k(x),$$

where
$$r_k(x) = s_k(x) + s_{2n+1-k}(x), \qquad 1 \le k \le n.$$

Theorem 5.4.15 (Chauhan, [75]). *For each n, $R_n(C[-1,1]) \subset \Pi_{4n+1}$. There exists a constant C such that, for each $f \in C[-1,1]$ and $x \in [-1,1]$,*

$$R_n(f, x_{kn}) = f(x_{kn}), \qquad 1 \le k \le n,$$

and

$$| f(x) - R_n(f,x) | \le C\omega \, (f, \Delta_n(x)) \, .$$

In 1998, He, Jiaxing and Li, Xiaoniu constructed another sequence with zeros of the Chebyshev polynomial of the second kind [184]. They were able to obtain an estimate in terms of the modulus of smoothness of order r, where r is an odd natural number. In [414] Xue-gang and De-hui constructed sequences based on the zeros of Jacobi polynomials and analyzed the rate of convergence.

5.5 Integral operators

We have presented in Section 2.5 some integral constructions due to Dzyadyk [109].

Taking into account the well-developed theory in trigonometric approximation, it is natural to ask whether one can obtain Jackson's theorem by considering convolution with algebraic polynomials. In the simplest case one can consider convolution with non-negative algebraic polynomials in order to obtain positive linear operators.

In 1963, Butzer raised the question of whether it is possible to construct polynomial operators, by means of singular convolution integrals, which approximate a function $f \in \mathrm{Lip}_\alpha[-1,1]$ with order $\mathcal{O}(n^{-\alpha})$, $0 < \alpha \le 1$. It was known that such an order of convergence can not be obtained with the usual changes in trigonometric operators. A natural extension of this problem is the following: construct a sequence of operators $L_n : C[-1,1] \to \mathbb{P}_n$ such that, if $C^1[-1,1]$ and $f' \in \mathrm{Lip}_\alpha[-1,1]$ $(0 < \alpha \le 1)$, then $\|f - L_n(f)\| = \mathcal{O}(n^{-1-\alpha})$.

In previous sections we presented some solutions of these problems. Here we only consider operators constructed by means of convolution.

For example, the Landau polynomials [224] are defined by

$$L_n(f,x) = C_n \int_{-1}^{1} f(t)(1-(t-x)^2)^n dt, \qquad 1/C_n = \int_{-1}^{1} (1-(t-x)^2)^n dt.$$

For $f \in C[-1, 1]$, $L_n(f)$ converges uniformly to f, but only on each interval $[-\delta, \delta]$, with $0 < \delta < 1$.

In 1968, DeVore proposed some operators obtained by convolution with the Legendre polynomials (3.5). Set

$$\Lambda_n(t) = C_n \frac{(P_{2n}(t))^2}{t^2 - x_{n+1}^2},$$

where P_{2n} is the Legendre polynomial of degree $2n$, x_{n+1} is the smallest positive zero of P_{2n} and C_n is chosen from the condition

$$\int_{-1}^{1} \Lambda_n(t)dt = 1.$$

Now, for each function $f \in C[-1/2, 1/2]$, define

$$L_n(f, x) = \int_{-1/2}^{1/2} f(t)\Lambda_n(t - x)dt. \tag{5.25}$$

Theorem 5.5.1 (DeVore, [86]). *For each $n \in \mathbb{N}$, let L_n be defined by (5.25).*

(i) *For each n, $L_n(C[-1/2, 1/2]) \subset \Pi_{4n-4}$.*
(ii) *If $f \in C[-1/2, 1/2]$ and $f(-1/2) = f(1/2) = 0$, then*

$$\|f - L_n(f)\| \le 40\,\omega(f, 1/n).$$

(iii) *If $M_n(x, f) = L(f, x) + L_n(f - Lf, x)$, where $L(f)$ is defined by (5.1) with $[a, b] = [-1/2, 1/2]$, then there exists a constant C such that, for $f \in C[-1/2, 1/2]$,*

$$\|f - M_n(f)\| \le C\,\omega(f, 1/n).$$

We remark that, using the properties of the first modulus of continuity and (ii), the constant C in (iii) can be taken as 120.

In 1969, Bojanic [32] showed that the last result holds for a large class of orthogonal polynomials.

Fix δ, c ($0 < \delta, c \le 1$). Let w be an even weight on $[-1, 1]$ satisfying the following conditions: there exist constant m and M such that

$$\begin{aligned} 0 < m \le w(x), \qquad & x \in [-c, c], \\ w(x) \le M, \qquad & x \in [-\delta, \delta]. \end{aligned}$$

Let $\{P_n\}$ be a family of orthogonal polynomials with respect to w and $-1 < x_{1,n} < \cdots < x_{n,n}$ be the zeros of P_n. That is

$$\int_{-1}^{1} P_n(x)x^k dx = 0, \quad 0 \le k \le n - 1 \quad \text{and} \quad \int_{-1}^{1} (P_n(x))^2 dx \ne 0.$$

Let $\{R_n\}$ be a sequence of polynomials defined by

$$R_n(x) = c_n \left(\frac{P_{2n}(x)}{x^2 - \alpha_{2n}^2} \right)^2 \quad \text{or} \quad R_n(x) = c_n \left(\frac{P_{2n+1}(x)}{x(x^2 - \alpha_{2n+1}^2)} \right)^2,$$

where α_j is the smallest positive zero of P_j and c_n is taken from the condition $\int_{-c}^{c} R_n(t)dt = 1$. For $f \in C[-c/2, c/2]$ and $n \in \mathbb{N}$ define

$$K_n(f, x) = \int_{-c/2}^{c/2} f(t) R_n(x - t) dt. \tag{5.26}$$

Theorem 5.5.2 (Bojanic, [32]). *For each $n \in \mathbb{N}$, let K_n be defined by (5.26). There exist constants C and N such that, for $f \in C[-c/2, c/2]$ and $n \geq N$,*

$$\|f - K_n(f)\| \leq C\,\omega(f, 1/n).$$

The proof is based on properties of the zeros of orthogonal polynomials, the Cotes numbers for gaussian quadratures and usual techniques for positive linear operators.

When $w(x) = 1/\sqrt{1 - x^2}$, the simplest case of the last theorem is obtained (this gives place to the Chebyshev polynomials) or $w(x) = \sqrt{1 - x^2}$ (this gives place to the Chebyshev polynomials of second kind). In the case $w(x) = 1$, we obtain the Legendre polynomials and we recover the DeVore theorem.

For the case of Chebyshev polynomials, a simplified proof was given by Bojanic and DeVore in 1969.

Theorem 5.5.3 (Bojanic-DeVore, [34]). *Let K_n be defined by (5.26) with $c = 1$ and $w(x) = 1/\sqrt{1 - x^2}$; for $f \in C[-1/2, 1/2]$ and $x \in [-1/2, 1/2]$, define*

$$K_n^*(f, x) = f(0) + K_n(f - f(0), x).$$

Then, for $n \geq 3$,

$$\|f - K_n^*(f)\|_{[-1/4, 1/4]} \leq 4\,\omega(f, 1/n).$$

In 1970, Chawla presented another construction based in convolution with an even positive polynomial.

Let $Q_n \in \mathbb{P}_n$ be an even polynomial, non-negative for $x \in [-1, 1]$. If $c_n = \int_{-1}^{1} Q_n(s)ds > 0$, define $P_n(x) = Q_n(x)/c_n$.

For $f \in C[-1/2, 1/2]$, Chawla considered the operator

$$L_n(f, Q_n, x) = \int_{-1/2}^{1/2} f(t) P_n(t - x) dt. \tag{5.27}$$

Theorem 5.5.4 (Chawla, [76]). *Let L_n be defined by (5.27). Assume*

$$f \in C[-1/2, 1/2] \qquad \text{and} \qquad f(-1/2) = f(1/2) = 0.$$

For $\delta > 0$ and $x \in [-1/2, 1/2]$, one has

$$\mid f(x) - L_n(f, Q_n, x) \mid \leq \left(\frac{3}{2} + \frac{\beta_n}{\delta}\right) \omega(f, \delta),$$

where $\beta_n^2 = (1 + \rho_{n,1})/(1 - \rho_{n,1})$ and $\rho_{n,1}$ is the coefficient of T_2 in the expansion of Q_n in terms of Chebyshev polynomials.

The best choice of Q_n leads to $\beta_n = \tan(\pi/(n+4))$. With this selection he proved

$$\mid f(x) - L_n(f, Q_n, x) \mid \leq 5\omega(f, 1/(n+4)).$$

Set $w(x) = 1/\sqrt{1-x^2}$ and consider $X = C[-1,1]$ or $X = L_p(w)$, $1 \leq p < \infty$. In 1976, Butzer and Stens studied the convergence properties of the singular integrals

$$I_\rho(f, x) = \frac{1}{\pi} \int_{-1}^{1} (\tau_x f)(u) \chi_\rho(u) w(u) du, \quad f \in X, \rho \in A \qquad (5.28)$$

where

$$\chi_\rho \in L_1(w), \qquad [\chi_\rho]^\wedge(0) = 1, \quad (\rho \in A),$$

$[\chi_\rho]^\wedge(k)$ is defined by (3.10) and $\tau_x f$ is the generalized translation given in (3.11) (see [56]). In [55] they estimated the rate of convergence.

Theorem 5.5.5 (Butzer and Stens, [55]). *Let $X = C[-1,1]$ or $X = L_p(w)$, $1 \leq p < \infty$. If the kernel $\{\chi_\rho\}_{\rho \in A}$ of the integral singular (5.28) is positive, then for all $f \in X$,*

$$\|f - I_\rho(f)\|_X \leq \left(1 + \frac{\pi}{\sqrt{2}}\right)^2 \omega_1^T \left(f, \cos\sqrt{1 - [\chi_\rho]^\wedge(1)}\right)_X,$$

where $\omega_1^T(f, t)_X$ is defined by (3.15).

Proof. We need the following inequality

$$\omega_1^T(f, \eta)_X \leq \left(1 + \frac{\arccos\eta}{\arccos\gamma}\right)^2 \omega_1^T(f, \gamma)_X, \qquad \text{for} \quad \gamma \in [-1, 1).$$

It can be obtained from the properties of the classical modulus of continuity. In fact,

$$\omega_1^T(f, \eta)_X = \omega_2(f \circ \cos, \arccos(\eta))_X$$
$$\leq \left(1 + \frac{\arccos\eta}{\arccos\gamma}\right)^2 \omega_2(f \circ \cos, \arccos(\gamma))_X$$
$$= \left(1 + \frac{\arccos\eta}{\arccos\gamma}\right)^2 \omega_1^T(f, \eta)_X.$$

We also need some estimates for the moments. Taking into account that $\arccos^2 u \le (\pi^2/2)(1 - u)$ for $u \in [-1, 1]$, one has

$$\frac{1}{\pi} \int_{-1}^{1} (\arccos u)^2 \chi_\rho(u) w(u) du \le \frac{\pi^2}{2} \frac{1}{\pi} \int_{-1}^{1} (1 - u) \chi_\rho(u) w(u) du = \frac{\pi^2}{2} (1 - [\chi_\rho]^\wedge(1)).$$

On the other hand, using Hölder's inequality we obtain

$$\frac{1}{\pi} \int_{-1}^{1} (\arccos u) \chi_\rho(u) w(u) du \le \sqrt{\frac{1}{\pi} \int_{-1}^{1} (\arccos u)^2 \chi_\rho(u) w(u) du \frac{1}{\pi} \int_{-1}^{1} \chi_\rho(u) w(u) du}$$

$$\le \frac{\pi}{\sqrt{2}} \sqrt{1 - [\chi_\rho]^\wedge(1)}.$$

Notice that $1 - [\chi_\rho]^\wedge(1) > 0$. Because if we assume

$$0 = 1 - [\chi_\rho]^\wedge(1) = \frac{1}{\pi} \int_{-1}^{1} (1 - u) \chi_\rho(u) w(u) du,$$

then $\chi_\rho(u) = 0$ (a.e.) and this is a contradiction ($[\chi_\rho]^\wedge(0) = 1$).
Finally, for each $\lambda \in [-1, 1)$,

$$\|f - I_\rho(f)\|_X \le \frac{1}{\pi} \int_{-1}^{1} \|(\tau_u f)(\cdot) - f(\cdot)\|_X \chi_\rho(u) w(u) du$$

$$\le \frac{1}{\pi} \int_{-1}^{1} \omega_1^T(f, u)_X \chi_\rho(u) w(u) du$$

$$\le \omega_1^T(f, \lambda)_X \frac{1}{\pi} \int_{-1}^{1} \left(1 + \frac{\arccos u}{\arccos \lambda}\right)^2 \chi_\rho(u) w(u) du$$

$$\le \omega_1^T(f, \lambda)_X \left(1 + \frac{\pi}{\sqrt{2} \arccos \lambda} \sqrt{1 - [\chi_\rho]^\wedge(1)}\right)^2.$$

Hence, the assertion follows by taking $\lambda = \cos(\sqrt{1 - [\chi_\rho]^\wedge(1)})$. □

In particular, this theorem can be used to estimate the rate of convergence of Fejér's means with respect to the Chebyshev systems. That is

$$\sigma_n(f, x) = (f * F_n)(x) \tag{5.29}$$

where

$$F_n(x) = 1 + 2 \sum_{k=1}^{n} \left(1 - \frac{k}{n+1}\right) T_k(x).$$

Corollary 5.5.6. *Let $X = C[-1, 1]$ or $X = L_p(w)$, $1 \le p < \infty$. For the Fejér singular integral (5.29) one has*

$$\|f - \sigma_n(f)\|_X \le \left(1 + \frac{\pi}{\sqrt{2}}\right)^2 \omega_1^T \left(f, \cos \frac{1}{\sqrt{1+n}}\right)_X.$$

The Fejér-Korovkin sums are defined by

$$K_n(f, x) = (f * \kappa_n)(x)$$

where

$$\kappa_n(x) = 1 + 2 \sum_{k=1}^{n} \mu_n(k) T_k(x)$$

and

$$\mu_n(k) = \frac{(n - k + 3) \sin \frac{(k+1)\pi}{n+2} - (n - k + 1) \sin \frac{(k-1)\pi}{n+2}}{2(n+2) \sin(\pi/(n+2))}.$$

Corollary 5.5.7. *For the Fejér-Korovkin sums one has, for $f \in X$,*

$$\|f - K_n(f)\|_X \leq \left(1 + \frac{\pi}{\sqrt{2}}\right)^2 \omega_1^T \left(f, \cos \sqrt{1 - \cos \frac{\pi}{n+2}}\right)_X.$$

These operators provide polynomials of degree n.

Interpolatory operators usually use zeros of orthogonal polynomials. These ideas can not be used for equidistant nodes, because they have a very bad behavior in interpolatory processes.

Szabados [362] constructed some operators of the form

$$L_n(f, x) = \sum_{k=0}^{n} P_{k,n}(x) f\left(\frac{k}{n}\right),$$

where $P_{k,n} \in \mathbb{N}$ and

$$\|f - L_n(f)\| \leq C\omega\left(f, \frac{1}{n}\right),$$

for $f \in C[-1, 1]$. He also constructed operators of the form

$$S_{n,r}(f, x) = \sum_{k=0}^{n,o} \sum_{j=0}^{r} P_{j,k,r,n}(x) f^{(j)}\left(\frac{k}{n}\right), \qquad (5.30)$$

for a family of polynomials $P_{j,k,r,n} \in \mathbb{P}_n$ and $f \in C^r[-1, 1]$.

Szabados asked if it is possible to improve the Jackson order $n^{-r}\omega(f^{(r)}; n^{-1})$, by means operators like in (5.30), to the Timan-Teliakovskii point-wise estimate $(\sqrt{1-x^2}/n)^r \omega(f^{(r)}; \sqrt{1-x^2}/n)$. In [398] Vértesi answered the question in the negative.

The operators defined by DeVore [86], Bojanic-DeVore [34], Bojanic [32], Szabados [362] (1976) and Butzer-Stens [55] are of degree $4n - 2$ or $4n - 4$ and some approximate f uniformly only on $[-1 + \varepsilon, 1 - \varepsilon]$ for each $\varepsilon > 0$.

Bavinck [19], Lupaş [242] and Stens-Wehrens [350] considered the integral

$$J_{2n}(f, x) = \frac{1}{2} \int_{-1}^{1} f(u) \chi_{2n}(x, u) du \qquad (5.31)$$

where the kernel is given by

$$\chi_{2n}(x,u) = \frac{3}{n^2+3n+3} \sum_{k=0}^{2n} \frac{2k+1}{2} P_k(x)P_k(u) \int_{-1}^{1} P_k(t)[P_n^{(2,0)}(t)]^2 dt,$$

$P_n^{(\alpha,\beta)}$ being the Jacobi polynomial. Note that $\chi_{2n}(x,u) \geq 0$ and

$$\int_{-1}^{1} \chi_{2n}(x,u)du = 2.$$

Theorem 5.5.8 (Lupaş, [242]). *If J_{2n} is defined by (5.31), then for each $f \in C[-1,1]$, $J_{2n}(f) \in \mathbb{P}_{2n}$ and*

$$\|J_{2n}f - f\| \leq (1+2\sqrt{3})\omega_1(f,1/2n).$$

In [58] Butzer, Stens and Wehrens presented a systematic approach to study direct approximation theorems by algebraic convolution operators. They used the so-called Legendre transform method. The ideas are similar to one presented in Section 3.2, but using expansions in terms of the Legendre polynomials.

Let X stand either for the space $C[-1,1]$ or $L^p(-1,1) = L^p$, $1 \leq p < \infty$, of all real-valued measurable functions f defined on $[-1,1]$ for which the norm

$$\|f\|_p = \left(\frac{1}{2}\int_{-1}^{1} |f(u)|^p \, du\right)^{1/p}$$

is finite.

Let P_n be the Legendre polynomials (3.5). The Legendre transform of $f \in X$ is defined by

$$f^\wedge(k) = \int_{-1}^{1} f(u)P_k(u)du. \tag{5.32}$$

It can be proved that (5.32) defines a bounded linear operator mapping X into (c_0), the space of all real sequences $\{a_k\}_{k=0}^{\infty}$ such that $\lim_{k\to\infty} a_k = 0$.

The translation operator is defined in this setting by

$$(\tau_h f)(x) = \frac{1}{\pi} \int_{-1}^{1} f\left(xh + u\sqrt{1-x^2}\sqrt{1-h^2}\right) \frac{du}{\sqrt{1-u^2}}, \qquad (x,h \in [-1,1]).$$

For each $h \in [-1,1]$, τ_h defines a positive linear operator from X into itself with $\|\tau_h\|_{[X,X]} = 1$ and, for all $f \in X$,

$$\lim_{h\to 1-} \|\tau_h(f) - f\|_X = 0.$$

The modulus of continuity and Lipschitz class are defined as follows:

$$\omega_1^L(f,t) = \sup_{t \leq h \leq 1} \|\tau_h f - f\|_X, \qquad (t \in (-1,1)) \tag{5.33}$$

$$\text{Lip}_1^L(\alpha, X) = \{f \in X : \omega_1^L(f,t) = \mathcal{O}((1-t)^\alpha)\}. \tag{5.34}$$

Butzer, Stens and Wehrens studied conditions upon the sequence of functions $\{\chi_n\}_{n\in\mathbb{N}_0} \subset L^1(-1,1)$ such that

$$\lim_{n\to\infty} \|f * \chi_n - f\|_X = 0, \tag{5.35}$$

in order to investigate the rate of convergence in (5.35), expressing it in terms of the modulus of continuity (5.33). Results related with the Fejér means, the Fejér-Korovkin means, the Rogosinski means, and the de La Vallée-Poussin means (among others) were given, where all means are considered with respect to the Legendre expansion.

Theorem 5.5.9. *Let $\{\chi_\rho\}_{\rho\in A}$ be a positive kernel, and let φ be a strictly positive function defined on A such that $\lim_{\rho\to\rho_0} \varphi(\rho) = 0$. The following assertions are equivalent.*

(i) $| 1 - \chi^{\wedge}_\rho | = \mathcal{O}(\varphi(\rho)), \qquad \rho \to \rho_0.$

(ii) $\|f - I_\rho(f)\|_X \leq M\omega^L_1(f, 1 - \varphi(\rho))_X.$

Here we present only one application.

Corollary 5.5.10. *Let the Fejér-Legendre means be defined by*

$$\sigma_n(f, x) = \sum_{k=0}^{n} \left(1 - \frac{k}{n+1}\right) (2k + 1) f^{\wedge}(k)P_k(x).$$

If $f \in \mathrm{Lip}_1(\alpha, C[-1, 1])$, $(0 < \alpha < 1)$ (see (5.34)), then $\|\sigma_n f - f\|_C = \mathcal{O}(n^{-\alpha})$.

Let $C_w[-1, 1]$ be the class of all $f \in C[-1, 1]$ for which there exists a sequence $\{P_n\}$, $P_n \in \mathbb{P}_n$ such that

$$| f(x) - P_n(x) | \leq w\left(\frac{\sqrt{1 - x^2}}{n} + \frac{1}{n}\right),$$

where w is an increasing continuous function such that $w(0) = 0$ and $\omega(t_1 + t_2) \leq M(w(t_1) + w(t_2))$, for a fixed constant M.

Let $_n(f)$ be the partial sums of the Chebyshev-Fourier series of f and

$$\sigma_{n,m}(f, x) = \frac{1}{m+1} \sum_{k=n-m}^{n} S_k(f, x)$$

be de la Vallée-Poussin sums.

Theorem 5.5.11 (Omataev, [274]). *Fix $\theta \in (0, 1)$ and, for each $n \in \mathbb{N}$ fix $m \in \mathbb{N}$ such that $m \leq \theta n$. For each $f \in C_w[1, 1]$ one has*

$$| f(x) - S_{n,m}(f.x) | \leq C \left(\ln \frac{n}{m+1} + \mathcal{O}(1)\right)$$

$$\times \sum_{k=n-m}^{n} w \left(\frac{\sqrt{1 - x^2}}{k - n + m + 1} + \frac{1}{(k - n + 1 - 1)^2}\right).$$

Omataev obtained a similar theorem for the Chebyshev polynomials of second type. Other similar results were given by Labunetz [223].

We finish this section by presenting a sequence due to Lupaş. He considered the Chebyshev coefficients defined by

$$a_k(f) = \frac{2}{\pi} \int_{-1}^{1} f(t) T_k(t) \frac{dt}{\sqrt{1-t^2}}, \quad k \geq 0.$$

Define

$$\varphi_n(x) = a_n \frac{1 + T_{n+2}(x)}{(x - \cos(\pi/(n+2)))^2}, \quad a_n = \frac{1}{\pi(n+2)} \sin^2 \frac{\pi}{n+2}.$$

Notice that $\varphi_n \in \mathbb{P}_n$. Let $t_{k,n}$ be the Fourier-Chebyshev coefficients of φ_n. That is

$$\varphi_n(x) = t_{0,n} + \sum_{k=1}^{n} t_{k,n} T_k(x).$$

Now define a kernel $L_n : [-1,1] \times [-1,1] \to \mathbb{R}$ by

$$L_n(x,t) = \sum_{k=0}^{n} t_{k,n} T_k(x) T_k(t).$$

It can be proved that $L_n(x,t) \geq 0$, for $(x,t) \in [-1,1] \times [-1,1]$. Thus the linear operator $J_n : C[-1,1] \to \mathbb{P}_n$ defined by

$$J_n(f,x) = \int_{-1}^{1} L_n(x,t) f(t) dt$$

is positive.

Theorem 5.5.12 (Lupaş, [243]). For $f \in C[-1,1]$,

$$| f(x) - J_n(f,x) | \leq \left(1 + \pi\sqrt{2} + \pi^2/2\right) \omega(f, \Delta_n(x)),$$

and

$$\|f - J_n(f)\| \leq 8\omega(f, 1/(n+2)).$$

Proof. First, notice that

$$\varphi_n(x_{kn}) = \begin{cases} (n+2)/2\pi, & k = 1, \\ 0, & 2 \leq k \leq n, \end{cases}$$

and

$$\varphi_n(-1) = \frac{1 + (-1)^n}{\pi(n+2)} \sin^2 \frac{\pi}{2(n+2)}.$$

We will use a quadrature formula (see Lemma 1 in [243]: if $g \in C^{n+2}[-1, 1]$, there exists $\theta = \theta(g, n)$, $\theta \in (-1, 1)$, such that

$$\int_{-1}^{1} \frac{g(t)}{\sqrt{1 - t^2}} dt = \frac{2\pi}{n + 2} \left(\frac{1 - (-1)^n}{4} g(-1) + \sum_{k=1}^{s} g(x_{kn}) \right) + R_n(g),$$

where $s = 1[n/2]$,

$$R_n(g) = \frac{\pi}{2^{n+1}} \frac{g^{(n+2)}(\theta)}{(n + 2)!} \quad \text{and} \quad x_{kn} = \frac{(2k - 1)\pi}{n + 2}.$$

Notice that,

$$\int_{-1}^{1} \varphi_n(t) w(t) dt = 1,$$

$$\int_{-1}^{1} \varphi_n(t)(1 - t) w(t) dt = 2 \sin^2 \frac{\pi}{2(n + 2)},$$

and

$$\int_{-1}^{1} \varphi_n(t)(1 - t)^2 w(t) dt = \frac{n + 1}{n + 2} \sin^2 \frac{\pi}{n + 2}.$$

Moreover

$$\int_{-1}^{1} \frac{\varphi_n(t)}{\sqrt{1 + t}} dt = \int_{-1}^{1} \varphi_n(t) \sqrt{1 - t} w(t) dt$$

$$\leq \sqrt{\int_{-1}^{1} \varphi_n(t) w(t) dt \int_{-1}^{1} \varphi_n(t)(1 - t) w(t) dt}$$

$$= \sqrt{2 \sin^2 \frac{\pi}{2(n + 2)}} < \frac{\pi \sqrt{2}}{2n}$$

and, we similar arguments we obtain

$$\int_{-1}^{1} \varphi_n(t) dt = \int_{-1}^{1} \varphi_n(t) \sqrt{1 - t^2} w(t) dt < \frac{\pi}{n}.$$

Set $z_1(t, x) =| x - tx - \varphi(x)\varphi(t) |$, $z_2(t, x) =| x - tx + \varphi(x)\varphi(t) |$ and $Q_n(t) = 2n\sqrt{1 - t} + n^2(1 - t)$. It can be proved that, for $t, x \in [-1, 1]$

$$z_j(t, x) \leq \Delta_n(x) Q_n(t), \qquad j = 1, 2.$$

The estimates given above yield

$$k_n = 1 + \int_{-1}^{1} \varphi_n(t) Q_n(t) w(t) dt < 1 + \sqrt{2}\pi + \pi^2/2.$$

On the other, hand, if $x, t \in [-1, 1]$, then

$$| f(x) - (\tau_x f)(t) | \leq \frac{1}{2} | f(x) - f(xt + \sqrt{1 - x^2}\sqrt{1 - t^2}) |$$
$$+ \frac{1}{2} | f(x) - f(xt - \sqrt{1 - x^2}\sqrt{1 - t^2}) |$$
$$\leq \frac{1}{2}\omega(f, z_1(t, x)) + \frac{1}{2}\omega(f, z_2(t, x))$$
$$\leq \omega(f, Q_n(t)\Delta_n(x))$$
$$\leq (1 + Q_n(t))\omega(f, \Delta_n(x)).$$

Finally, one has

$$| f(x) - J_n(f, x) | \leq \int_{-1}^{1} | f(x) - (\tau_x f)(t) | \varphi_n(t)w(t)dt$$
$$\leq \omega(f, \Delta_n(x)) \int_{-1}^{1} (1 + Q_n(t))\varphi_n(t)w(t)dt$$
$$\leq (1 + \sqrt{2}\pi + \pi^2/2)\, \omega(f, \Delta_n(x)). \qquad \square$$

Define

$$J_n^*(f, x) = J_n(f, x) + (1 - x)/2[f(-1) - J_n(f, -1)] + (1 + x)/2[f(1) - J_n(f, 1)].$$

Theorem 5.5.13 ([243]). *There exists a constant C such that, for $f \in C[-1, 1]$,*

$$| f(x) - J_n^*(f, x) | \leq C\, \omega(f, \sqrt{1 - x^2}/n).$$

In 1989, Shevchuk [339] found a simple representation of Dzyadyk's polynomial kernel in connection with the segment: $[-1, 1]$. For $r, n \in \mathbb{N}$ let $J_{n,r}$ be the Jackson kernel

$$J_{n,r}(t) = \frac{1}{\gamma_{n,r}} \left(\frac{\sin nt/2}{\sin t/2} \right)^{2(r+1)}$$

where $\gamma_{n,r}$ is chosen from the conditions $\int_{-\pi}^{\pi} J_{n,r}(t)dt = 1$.

Now for $m, n, r \in \mathbb{N}$, $x, y \in [-1, 1]$; $\beta = \arccos x$ and $\alpha = \arccos y$ set

$$D_{m,n,r}(y, x) = \frac{1}{(m-1)!} \frac{\partial^m}{\partial x^m} (x - y)^{m-1} \int_{\beta-\alpha}^{\beta+\alpha} J_{n,r}(t)dt.$$

The kernel $D_{m,n,r}(y, x)$ is an algebraic polynomial of degree $(r + 1)(n - 1) - 1$ with respect to the variable x and $D_{m,1,r}(y, x) = 0$.

The idea for such a representation is based on the function

$$\varphi_{n,r}(x, y) = \int_{\beta-\alpha}^{\beta+\alpha} J_{n,r}(t)dt.$$

used by DeVore in [91]. With this kernel we define the operator

$$L_n(f,x) = \int_{-1}^{1} f(y)D_{m,n,r}(y,x)dy.$$

It can be verified that $L_n(f,x)$ is an algebraic polynomial and the sequence $\{L_n(f,x)\}$ approximates the function f and its derivatives.

In [113] the result of Shevchuk was presented in a more general form, which included the estimate of Ditzian-Totik and Trigub simultaneous approximation type theorems. They used a variant of Dzyadyk's kernel that was developed by Shevchuk in [339] for complex approximation. In [173] he showed that the Ditzian-Totik and the τ modulus are equivalent.

5.6 Simultaneous approximation

From the Trigub and Gopengauz result we know that certain interpolation processes can be used for simultaneous approximation.

The ideas of Gopengauz were used by Baiguzov [7] to obtain results in approximate differentiation with the aid of Lagrange interpolatory polynomials.

In 1981, Srivastava studied the first derivatives of the operators constructed by Kis and Vértesi [199]. In fact, he modified the operators in order to estimate the first derivative. Let $-1 \le x \le 1$, $x = \cos t$,

$$x_{k,n} = \cos t_{k,n}, \qquad t_{k,n} = \frac{2k\pi}{2n+1}, \qquad k = 0,\dots,n.$$

For $k = -n,\dots,n$, define

$$l_{k,n}(t) = \frac{\sin(2n+1)(t - t_{k,n}/2)}{(2n+1)\sin(t - t_{k,n})/2}. \tag{5.36}$$

Then for $f \in C^s[-1,1]$ we define the polynomial

$$L_{n,s}(f,x) = \sum_{k=0}^{n}\sum_{r=0}^{s}(x - x_{k,n})^r f^{(r)}(x_{k,n})v_k(x), \qquad (s = 0,1) \tag{5.37}$$

where

$$v_0(x) = u_0(t), \qquad v_k(x) = u_k(t) + u_{-k}(t), \qquad (1 \le k \le n),$$
$$u_k(t) = 4l_k^3(t) + 3l_k^4(t),$$

$k = -n,\dots,n$. For $s = 0$, this polynomial is the same as the one of Kis and Vértesi.

Theorem 5.6.1 (Srivastava, [344]). *Let $L_{n,1}$ be given by (5.37) (with $s = 1$). For $f \in C[-1, 1]$, $n \in \mathbb{N}$ and $x \in [-1, 1]$ one has*

$$| f(x) - L_{n,1}(f, x) | \le \frac{C_1}{n} \omega \left(f, \frac{1}{n} \right),$$

and

$$| f'(x) - L'_{n,1}(f, x) | \le C_2 \omega \left(f, \frac{1}{n} \right).$$

In 1978, Vértesi constructed linear polynomial operators of degree $\le 2n(1+c)$ which interpolate f and f' at the Chebyshev nodes (assuming f' is continuous). Moreover, he provided Teliakovskii-Gopengauz-type estimates.

With $t_{k,n} = \cos(2k-1)\pi/(2n)$ and $x_{k,n} = \cos t_{k,n}$, $k = 1, 2, \ldots, n$, set

$$l_{k,n}(x) = \frac{(-1)^{k+1} \sin t_{k,n} T_n(x)}{n(x - t_{k,n})}, \qquad v_{k,n}(x) = \frac{1 - xx_{k,n}}{1 - x_{k,n}^2},$$

$$h_{k,n}(x) + v_{k,n}(x)l_{k,n}^2(x), \qquad G_{k,n}(x) = (x - x_{k,n})l_{k,n}^2(x).$$

Now, for $f \in C^1[-1, 1]$, define

$$H_n(f, x) = \sum_{k=1}^{n} f(x_{k,n})h_{k,n}(x) + \sum_{k=1}^{n} f'(x_{k,n})G_{k,n}(x).$$

Fejér [118] proved that

$$H_n(f, x_{k,n}) = f(x_{k,n}) \quad \text{and} \quad H_n'(f, x_{k,n}) = f'(x_{k,n}).$$

Furthermore, $H_n(f)$ converges uniformly to f. But the rate of convergence could be very slow.

For the new construction, fix $s = s(n) \le n$ and $n \le Cs$. Set

$$\min_{1 \le i \le s} | t_{k,n} - t_{i,s} | = | t_{k,n} - t_{j_k,s} |, \qquad (k = 1, \ldots, n),$$

(if there is more than one point satisfying this, choose any of them).

Define

$$F_{k,n}(x) = \frac{l_{j_k,s}^{r+2}(x) \sin^{2r+2}(t)}{l_{j_k,s}^{r+3}(x_{k,n}) \sin^{2r+2} t_{k,n}}$$

$$\times \left\{ h_{k,n}(x) + \left[(2r+2)\frac{\cos t_{k,n}}{\sin^2 t_{k,n}} - (r+3)\frac{l'_{j_k,s}(x_{k,n})}{l_{j_k,s}(x_{k,n})} \right] G_{k,n}(x) \right\}$$

and

$$D_{k,n}(x) = \frac{l_{j_k,s}^{r+3}(x) \sin^{2r+2} t}{l_{j_k,s}^{r+3}(x_{k,n}) \sin^{2r+2} t_{k,n}},$$

where $x = \cos t$.

Now, for $f \in C^r[-1, 1]$ ($r \geq 1$), define the operator

$$A_n(f, x) = L_{n,r}(f, x) + \sum_{k=1}^{n} [f(x_{k,n}) - L_{n,r}(f, x_{k,n})] F_{k,n}(x)$$

$$+ \sum_{k=1}^{n} [f'(x_{k,n}) - L'_{n,r}(f, x_{k,n})] D_{k,n}(x),$$

where $L_{n,r}$ is the operator of Gopengauz given in Theorem 2.8.11. It can be proved that, if

$$s = \left[\frac{2nc - r + 2}{r + 3} \right],$$

then for each $f \in C^r[-1, 1]$,

$$\deg A_n(f) \leq (r + 3)(s - 1) + 2r + 2 + 2n - 1 \leq 2n(1 + c).$$

Theorem 5.6.2 (Vértesi, [399]). *For every $c > 0$ fixed and $r \geq 1$, let A_n be the linear polynomial operators defined above. One has*

(i) $A_n(C^r[-1, 1]) \subset \Pi_{2n(1+c)}$,

(ii) $A_n(f, x_{k,n}) = f(x_{k,n})$ *and* $A'_n(f, x_{k,n}) = f'(x_{k,n})$, *for* $k = 1, 2, \ldots, n$ *and* $n \geq n_0$,

(iii) $| f^{(i)}(x) - A_n^{(i)}(x) | \leq C \left(\frac{\sqrt{1 - x^2}}{n} \right)^{r-1} \omega \left(f^{(r)}, \frac{\sqrt{1 - x^2}}{n} \right), (0 \leq i \leq r)$

for $n \geq n_0$ and $f \in C^r[-1, 1]$.

Vértesi considered also other operators. For $f \in C^r[-1, 1]$ and $n \geq n_0$ define

$$B_n(f, x) = L_{n,r}(f, x) + \sum_{k=1}^{n} [f(x_{k,n}) - L_{n,r}(f, x_{k,n})] F_{k,n}(x).$$

That is, the term containing the derivatives in A_n is omitted.

Theorem 5.6.3 ([399]). *For every $c > 0$ fixed and $r \geq 0$, consider the linear polynomial operators B_n defined above. One has:*

(i) $B_n(C^r[-1, 1]) \subset \Pi_{2n(1+c)}$, $n \geq n_0$,

(ii) $B_n(f, x_{k,n}) = f(x_{k,n})$ *and* $B'_n(f, x_{k,n}) = L'_{n,r}(f, x_{k,n})$, *for* $k = 1, 2, \ldots, n$ *and* $n \geq n_0$,

(iii) $| f^{(i)}(x) - B_n^{(i)}(x) | \leq C \left(\frac{\sqrt{1 - x^2}}{n} \right)^{r-i} \omega \left(f^{(r)}, \frac{\sqrt{1 - x^2}}{n} \right), (0 \leq i \leq r)$

for $n \geq n_0$ and $f \in C^r[-1, 1]$.

In [399] Vértesi also presented some estimates in terms of the best approximation and showed that the results of Saxena [325] (see Theorem 5.7.2) and Rodina
[313] (see Theorem 5.3.2) can be obtained from his approach.

In 1978 Saxena and Srivastava [327] proved some results considering interpolation of the function and its first derivative (see also [329]). One year later they
obtained another that we present here.

Let $l_{k,n}$ be given as in (5.36) and set

$$p_{kn}(x) = \frac{1}{43}[1008l_{k,n}^5(t) - 1820l_{k,n}^6(x) + 960l_{n,k}^7(t) - 105l_{k,n}^9(t).$$

Define $q_{0,0}(x) = p_{0,0}(t)$ and $q_{k,n,0}(x) = p_{k,n}(t) - p_{-k,n}(t)$, $1 \le k \le n$.
Now, for $f \in C^1[-1,1]$ consider the operators

$$Q_{n,0}(f,x) = L(f,x) + \sum_{k=0}^{n} (f(x_{k,n}) - L(f,x)q_{k,n}(x))$$

and

$$Q_{n,1}(f,x) = Q_{n,0}(f,x) + \sum_{k=0}^{n}(x - x_{k,n})f'(x_{k,n})q_{k,n}(x).$$

Theorem 5.6.4 (Saxena and Srivastava, [328]). *If $Q_{n,0}$ and $Q_{n,1}$ are defined as
above, then $Q_{n,i}(C[-1,1]) \subset \Pi_{8n+1}$ $(i = 0,1)$. There exists constants C_0 and C_1
such that, for $f \in C^1[-1,1]$, $Q_{n,0}(f)$ and $Q_{n,1}(f)$ interpolate f and f' at the
points $\{x_{k,n}\}$ respectively,*

$$| f(x) - Q_{n,0}(f,x) | \le C_0\tilde{\omega}(f,\Delta_n(x))$$

and

$$| f'(x) - Q_{n,1}(f,x) | \le C_1\Delta_n(x)\tilde{\omega}(f',\Delta_n(x)),$$

where $\tilde{\omega}$ denotes the least concave majorant of the modulus of continuity.

Later, in 1985, Gonska and Hinnemann showed that generalization to a
higher-order modulus is possible, if we consider linear operators defined for differentiable functions. It will be presented in the section devoted to boolean sums. By
extending the ideas of Gonska, Hinnemann and Yu, Dahlhaus proved that (2.38)
holds with a modulus of order r if and only if $0 \le k \le \min\{s - r + 2, s\}$.

Theorem 5.6.5 (Dahlhaus, [79]). *Let $r, s \in \mathbb{N}_0$. There exists a constant $C = C(r,s)$
such that, for all $f \in C^s[-1,1]$ and all $n \ge \{\max(4(s+1), r+s\}$, there exists
$P_n \in \mathbb{P}_n$ such that*

$$| f^{(k)}(x) - P_n^{(k)}(x) | \le C_s (\delta_n(x))^{s-k} \omega_r \left(f^{(s)}, \delta_n(x)\right),$$

for all $k \in \mathbb{N}_0$ with $0 \le k \le \min\{s - r + 2, s\}$ and all $x \in [-1,1]$.

Theorem 5.6.6 ([79]). *Let $r, s \in \mathbb{N}_0$. For all $C \in \mathbb{R}$ and all $n \in \mathbb{N}$, there exists a function $f \in C^k[-1, 1]$ such that, for all $P_n \in \mathbb{P}_n$, there exists an $x = x_k \in [-1, 1]$ such that*

$$| f^{(k)}(x) - P_n^{(k)}(x) |> C \left(\delta_n(x)\right)^{s-k} \omega_r \left(f^{(s)}, \delta_n(x) \right),$$

for all $k \in \mathbb{N}$, with $s - r + 3 \leq k \leq s$.

Li (independent of Dahlhaus) proved the following result and showed that the estimate is the best possible in some sense.

Theorem 5.6.7 (Li, [235]). *Fix $r \geq m + 2$. For any $n \geq r + m - 1$ there exists a linear operator $Q_n : C^m[-1, 1] \to \mathbb{P}_n$ such that, for $f \in C^m[-1, 1]$, $0 \leq k \leq m$ and $x \in [-1, 1]$,*

$$| f^{(k)}(x) - Q_n^{(k)}(f, x) |\leq C\delta_n^{m-k}(x)\omega_r \left(f^{(m)}, \delta_n(x) + \frac{(n\sqrt{1 - x^2})^{(m+2-k)/r}}{n^2} \right).$$

There exists a sequence of polynomials which converges to a differentiable function at the rate given in Timan's theorem and also interpolates the function on an array of points converging to ± 1 at a prescribed rate of $O(n^{-2})$. From the point of view of interpolation theory, Gopengauz-Teliakovskii-type theorems give polynomials which interpolate the derivatives $f^{(k)}$ $0 \leq k \leq m - 1$ at the points ± 1, a fact which has made this theorem useful in recent investigations of simultaneous approximation by interpolation. Balázs, Kilgore and Vértesi showed that the estimates for simultaneous approximation can be combined with certain interpolatory properties (see [11]. In particular, they considered interpolation at (not necessarily) distinct points clustered near ± 1.

Theorem 5.6.8 (Balázs-Kilgore-Vértesi, [12]). *Let $f \in C^q[-1, 1]$. Let $r = [(q + 1)/2]$, and let a constant $C > 0$ be given. Let points $t_{0,n}, \ldots, t_{r-1,n}$ and $s_{0,n}, \ldots, s_{r-1,n}$ be given such that for each $n \geq \max\{2r, C^{1/2}\}$,*

$$-1 \leq t_{0,n} \leq \cdots \leq t_{r-1,n} \leq -1 + C/n^2$$

and

$$1 \geq s_{0,n} \geq \cdots \geq s_{r-1,n} \geq 1 - C/n^2.$$

Then, for each such n there exists a polynomial P_n of degree n or less, such that for $| x |\leq 1$ and for $k = 0, \ldots, q$,

$$|f^{(k)}(x) - P_n^{(k)}(x)| \leq C \left(\frac{\sqrt{1 - x^2}}{n} + \frac{1}{n^2} \right)^{q-k} \omega \left(f^{(q)}; \frac{\sqrt{1 - x^2}}{n} + \frac{1}{n^2} \right),$$

or

$$|f^{(k)}(x) - P_n^{(k)}(x)| \leq C \left(\frac{\sqrt{1 - x^2}}{n} + \frac{1}{rn^2} \right)^{q-k} E_{n-q}(f^{(q)}),$$

and furthermore

$$P_n(x) = f(x), \quad x \in \{t_{0,n}, \dots, t_{r-1,n}, s_{0,n}, \dots, s_{r-1,n}\}.$$

If for any specific n there exist one (or more) j and l such that

$$t_{j,n} = t_{j+l,n} \cdots t_{j+l,n} \quad or \quad s_{j,n} = s_{j+1,n} = \cdots = s_{j+l,n},$$

then in addition

$$f^{(k)}(t_{j,n}) = P_n^{(k)}(t_{j,n}), \quad k = 0, \dots, l$$

or respectively

$$f^{(k)}(s_{j,n}) = P_n^{(k)}(s_{j,n}), \quad k = 0, \dots, l.$$

In Theorem 5.4.6 we present a work of Varma where he gave a new proof of the inequality of Brudnyi for the case $r = 1$. The process is of a weakly interpolatory type. It turns out the process developed in [395] cannot provide the proof of this inequality for $r = 2$. In [396] Varma and Yu presented another process.

For $k = 1, 2, \dots, n$, we denote by

$$l_{k,n}(x) = \frac{(-1)^{k+1}\sqrt{1 - x_{k,n}^2}}{n} \frac{T_n(x)}{x - x_{k,n}}$$

the fundamental polynomials of Lagrange interpolation based on the nodes $x_{k,n}$ where

$$x_{k,n} = \cos\frac{(2k-1)\pi}{2n}, \quad k = 1, \dots, n,$$

are the zeros of $T_n(x)$ in $(-1, 1)$.

Write

$$\psi_{1,n}(x) = \frac{1}{4}(3l_{1,n}(x) + l_{2,n}(x)), \quad \psi_{n,n}(x) = \frac{1}{4}(3l_{n-1,n}(x) + l_{n,n}(x)),$$

$$\psi_{k,n}(x) = \frac{1}{4}(l_{k-1,n}(x) + 2l_{k,n}(x) + l_{k+1,n}(x)), \quad k = 2, \dots, n-1,$$

$$\chi_{1,n}(x) = \frac{1}{4}(3\psi_{1,n}(x) + \psi_{2,n}(x)), \quad \chi_{n,n}(x) = \frac{1}{4}(3\psi_{n-1,n}(x) + \psi_{n,n}(x)),$$

and

$$\chi_{k,n}(x) = \frac{1}{4}(\psi_{k-1,n}(x) + 2\psi_{k,n}(x) + \psi_{k+1,n}(x)), \quad k = 2, \dots, n-1.$$

With this notation, for $f \in C[-1, 1]$ define

$$G_n(f, x) = \sum_{k=1}^{n} f(x_{k,n})\chi_{k,n}(x)$$

and

$$H_n(f, x) = G_n(f, x) - \frac{1+x}{2}(G_n(f, 1) - f(1)) - \frac{1-x}{2}(G_n(f, -1) - f(-1)).$$

Theorem 5.6.9 (Varma and Yu, [396]).

(i) *If* $f \in C[-1,1]$, *then*

$$| f(x) - H_n(f,x) | \le C \omega_2 \left(f, \frac{\sqrt{1-x^2}}{n} \right).$$

(ii) *If* $f \in C^1[-1,1]$, *then*

$$| f'(x) - H_n'(f,x) | \le C \omega \left(f, \frac{\sqrt{1-x^2}}{n} + \frac{1}{n^2} \right).$$

Furthermore if

$$R_n(f,x) = H_n'(f,x) - \frac{1+x}{2}(H_n'(f,1) - f(1)) - \frac{1-x}{2}(H_n'(f,-1) - f(-1)),$$

then

$$| f'(x) - R_n(f,x) | \le C \omega \left(f', \frac{\sqrt{1-x^2}}{n} \right).$$

The next theorem provides the solution of the problem of simultaneous approximation of a function and its derivatives through interpolation polynomials (weak interpolation).

Theorem 5.6.10 ([396]). *If* $f \in C^1[-1,1]$ *and*

$$J_n(f,x) = H_n(f,x) - \frac{(1+x)^2}{4n^2}(T_n(x) - T_n(1))(H_n'(f,1) - f(1))$$
$$- \frac{(1-x)^2(-1)^{n+1}}{4n^2}(T_n(x) - T_n(-1))(H_n'(f,-1) - f(-1)),$$

then

$$| f^{(r)}(x) - J_n^{(r)}(f,x) | \le C \frac{\sqrt{1-x^2}}{n} \omega \left(f', \frac{\sqrt{1-x^2}}{n} \right),$$

for $r = 0,1$.

Let $f \in C^q[-1,1]$ be given, where $q \ge 0$. Then for a fixed r such that $q/2 < r \le q+1$ we define a polynomial $H_{n,r}(f,x)$ of degree at most $n + 2r - 1$ which interpolates f on nodes $x_1, \ldots, x_n$ such that $-1 < x_n < \cdots < x_1 < 1$ and interpolates $f^{(0)}, \ldots, f^{(r-1)}$ at ± 1. The polynomial $H_{n,r}(f,x)$ may be represented as

$$H_{n,r}(f,x) = \sum_{j=1}^{n} f(x_j) \left(\frac{1-x^2}{1-x_j^2} \right)^r l_j(x) + \sum_{k=0}^{r-1} [f^{(k)}(1)h_{1,k}(x) + f^{(k)}(-1)h_{2,k}(x),$$

where

$$l_j(x) = \prod_{s=1, s\neq j}^{n} \frac{x - x_s}{x_j - x_s}$$

and $h_{1,k}(x)$ and $h_{2,k}(x)$ are certain polynomials of degree $n + 2r - 1$.

The approximation properties of $H_{n,r}$ are described in terms of the weighted Lebesgue sums

$$L_{n,s}(x) = \sum_{j=1}^{n} \left(\frac{1 - x^2}{1 - x_j^2} \right)^{s/2} |l_j(x)|.$$

We remark that $L_{n,0}$ is the ordinary Lebesgue sum of the Lagrange interpolation on the nodes $x_1, \ldots, x_n$.

Theorem 5.6.11 (Kilgore and Prestin, [194]). *Let $f \in C^q[-1,1]$. Then for $q/2 < r \leq q + 1$,*

$$|f(x) - H_{n,r}(f, x)| \leq M_q \left(\frac{\sqrt{1 - x^2}}{n} \right)^q w\left(f^{(q)}, \frac{\sqrt{1 - x^2}}{n} \right)$$
$$\times (1 + \max\{2L_{n,2r-q-1}(x), L_{n,2r-q}(x)\}).$$

A similar statement holds with w replaced by w_2:

$$|f(x) - H_{n,r}(f, x)| \leq C \left(\frac{\sqrt{1 - x^2}}{n} \right)^q w\left(f^{(q)}, \frac{\sqrt{1 - x^2}}{n} \right)$$
$$\times (1 + \max\{4L_{n,2r-q-2}(x), L_{n,2r-q}(x)\}).$$

Furthermore, for the derivatives one has

Theorem 5.6.12 ([194]). *Let $f \in C^q[-1,1]$. Then for $q/2 < r \leq q + 1$ and for $k = 0, \ldots, q$ there is a constant C_q depending only upon q such that*

$$|f^{(k)}(x) - H_{n,r}^{(k)}(f, x)| \leq C_q \left(\frac{\sqrt{1 - x^2}}{n} + \frac{1}{n^2} \right)^{q-k} w\left(f^{(q)}, \frac{\sqrt{1 - x^2}}{n} + \frac{1}{n^2} \right)$$
$$\times \max\{\|L_{n,2r-q-1}\|, \|L_{n,2r-q}\|\}.$$

Furthermore, for $0 \leq k < r$ we have

$$|f^{(k)}(x) - H_{n,r}^{(k)}(f, x)| \leq eC_q \left(\frac{\sqrt{1 - x^2}}{n} \right)^{r-k} \left(\frac{\sqrt{1 - x^2}}{n} + \frac{1}{n^2} \right)^{q-r}$$
$$\times w\left(f^{(q)}, \frac{\sqrt{1 - x^2}}{n} + \frac{1}{n^2} \right) \max\{\|L_{n,2r-q-1}\|, \|L_{n,2r-q}\|\}.$$

In the special case $r = q + 1$, there is a constant $K_q \leq \max\{4eC_q, 7C_q + 7\}$ such that for $k = 0, \ldots, q$,

$$| f^{(k)}(x) - H_{n,r}^{(k)}(f, x) | \leq K_q \left(\frac{\sqrt{1 - x^2}}{n} \right)^{q-k} w \left(f^{(q)}, \frac{\sqrt{1 - x^2}}{n} \right)$$

$$\times \max\{ \|L_{n,2r-q-1}\|, \|L_{n,2r-q}\| \}.$$

As a consequence of these theorems we can obtain point-wise estimates for the quality of approximation on Jacobi nodes with added interpolation at ± 1 which improve on what has been previously known by including the point-wise modulus of continuity or the point-wise modulus of smoothness.

Theorem 5.6.13 ([194]). *Let $f \in C^q[-1, 1]$ and r be given such that $q/2 < r \leq q+1$. Then for $2r - q - 5/2 \leq \alpha, \beta \leq 2r - q - 3/2$ we can choose the nodes x_j at the zeros of the ordinary Jacobi polynomials $P_n^{(\alpha, \beta)}$, and we obtain*

$$\max\{ 2L_{n,2r-q-1}(x), L_{n,2r-q}(x) \} \leq C \log n,$$

whence for these nodes

$$| f(x) - H_{n,r}(f, x) | \leq M_q \left(\frac{\sqrt{1 - x^2}}{n} \right)^q w \left(f^{(q)}, \frac{\sqrt{1 - x^2}}{n} \right) \log n,$$

in which the constant C depends upon q, α, β. Again using the nodes generated by $P_n^{(\alpha, \beta)}$ we obtain that

$$\max\{ 2L_{n,2r-q-2}(x), L_{n,2r-q}(x) \} \leq C \log n$$

if and only if $\alpha = \beta = 2r - q - 5/2$ so that for the nodes thus determined we obtain

$$| f(x) - H_{n,r}(f, x) | \leq M_q \left(\frac{\sqrt{1 - x^2}}{n} \right)^q w_2 \left(f^{(q)}, \frac{\sqrt{1 - x^2}}{n} \right) \log n.$$

Theorem 5.6.14 ([194]). *Let $f \in C^q[-1, 1]$ and $r = q + 1$. Then for $q - 1/2 \leq \alpha, \beta \leq q + 1/2$ we can choose the nodes x_j at the zeros of the ordinary Jacobi polynomials $P_n^{(\alpha, \beta)}$, and we obtain, for $k = 0, \ldots, q$,*

$$| f^{(k)}(x) - H_{n,r}^{(k)}(f, x) | \leq C \left(\frac{\sqrt{1 - x^2}}{n} \right)^{q-k} w \left(f^{(q)}, \frac{\sqrt{1 - x^2}}{n} \right) \log n,$$

in which the constant C depends upon q, α, β. Also, using again the nodes generated by the Jacobi polynomials, the statement

$$| f(x) - H_{n,r}(f, x) | \leq C \left(\frac{\sqrt{1 - x^2}}{n} \right)^q w_2 \left(f^{(q)}, \frac{\sqrt{1 - x^2}}{n} \right) \log n$$

holds for $\alpha = \beta = q - 1/2$.

Corollary 5.6.15. *There exists a sequence of linear operators* $Q_n : C^m[-1,1] \to \mathbb{P}_n$ *such that, for* $0 \leq k \leq m$ *and* $x \in [-1,1]$,

$$| f^{(k)}(x) - Q_n^{(k)}(x) | \leq C \left(\delta_n(x)\right)^{m-k} \omega_r \left(f^{(m)}, \Delta_n(x)\right).$$

In [8] and [9], Balász and Kilgore generalized ideas of Szabados [364] and Runck-Vértesi [317]. They considered interpolation by adding a certain set of points to the optimal ones.

Let $X_n = \{x_{1,n}, \ldots, x_{n,n}\}$ be a systems of nodes in $(-1,1)$. Set $r = [(q+1)/2]$ and choose another set of nodes $T_n = \{t_{0,n}, \ldots, t_{r-1}\} \cup \{s_{0,n}, \ldots, s_{r-1,n}\}$ satisfying the following conditions: for some $C > 0$ and an integer $N \geq \sqrt{C}$, for $0 \leq k \leq r$,

$$-1 \leq t_{k,n} \leq -1 + \frac{C}{(n+N)^2} < 1 - \frac{C}{(n+N)^2} \leq s_{n,k} \leq 1.$$

Nodes which lie upon the same point require Hermite interpolation.

Theorem 5.6.16 (Balász and Kilgore, [9]). *Fix* $q \in \mathbb{N}$ *and set* $r = [(q+1)/2]$. *Let* P_n *be the interpolation operators upon the nodes* $X_n \cup T_n$. *For* $f \in C^q[-1,1]$ *and* $x \in [-1,1]$ *one has:*

(i) *if* q *is even and* $0 \leq i \leq q$,

$$| f^{(i)}(x) - P_n^{(i)}(f,x) | \leq C \frac{1}{n^{q-i}} E_{n-1}(f^{(q)}) \|L_n\|,$$

(ii) *if* q *is odd and* $0 \leq i \leq q$,

$$| f^{(i)}(x) - P_n^{(i)}(f,x) | \leq C \frac{1}{n^{q-i}} E_{n-1}(f^{(q)}) \|L_n^*\|,$$

where L_n *is the Lagrange interpolation operators upon the nodes* X_n *and* $L_n^*(f,x) = \sqrt{1-x^2} L_n(\sqrt{1-t^2} f(t), x)$.

In [193] and [195] Kilgore and Prestin gave point-wise estimates and results of Gopengauz type. They used interpolation on Jacobi polynomials.

5.7 Estimation with constants

In 1970, Saxena modified his ideas in [323] to obtain the following result which provided an estimation for the constant in a Teliakovskii-type theorem.

Theorem 5.7.1 (Saxena, [324]). *For each* $f \in C[-1,1]$ *and* $n \in \mathbb{N}$ *there exists a linear operator* $L_{4n+2} : C[-1,1] \to \Pi_{4n+2}$ *such that*

$$| f(x) - P_{4n+2}(x) | \leq 384 \left(\omega\left(f, \frac{\sqrt{1-x^2}}{n}\right) + \omega\left(f, \frac{|x|}{n}\right)\right).$$

In 1972, Saxena used a similar construction to obtain Teliakovskii-type estimates. Set

$$l_{kn}(x) = \frac{(-1)^{k+n}(1 - x_{kn}^2)}{n+1} \frac{T_n(x)}{x - x_{kn}}, \qquad v_{kn}(x) = 1 - \frac{3x_k n}{x - x_{kn}} 1 - x_{kn}^2,$$

where T_n is the Chebyshev polynomials and x_{kn} are the zeros of T_n. Set

$$\psi_n(t, u) = \frac{2}{n+1} \sum_{r=1}^{n-1} T_r'(t) T_n(u)$$

and

$$\lambda_{kn}(x) = \left(\frac{1 - x^2}{1 - x_{kn}}\right)^2 \left[v_{kn}(x) l_{kn}^4(x) + 2(x - x_{kn} l_{kn}^3(x)(1 - x_{kn}^2)\psi_n(x_{kn}, x)\right].$$

Finally, define

$$L_n(f, x) = \frac{1 + x}{2} f(1) + \frac{1 - x}{2} f(-1)$$

$$+ \sum_{k=1}^{n} \left[f(x_{kn}) - \frac{1 + x}{2} f(1) - \frac{1 - x}{2} f(-1)\right] \lambda_{kn}(x).$$

Theorem 5.7.2 (Saxena, [325]). *If L_n is defined by the last equation, then for each $f \in C[-1, 1]$, $L_n(f) \in \Pi_{4n+2}$ and*

$$| f(x) - L_n(f, x) | \le 1285 \, \omega\left(f, \frac{\sqrt{1 - x^2}}{n}\right).$$

In 1978 Pichugov proved that some trigonometric kernels can be used to obtain polynomial operators.

Theorem 5.7.3 (Pichugov, [282]). *For arbitrary numbers $\rho_{1,n}$ and $\rho_{2,n}$, which are the coefficients of a positive trigonometric polynomial*

$$K_n(t) = \frac{1}{2} + \sum_{k=1}^{n} \rho_{k,n} \cos(kt),$$

there exists a linear operator $L_n : C[-1, 1] \to \mathbb{P}_{n-1}$ such that, for all $f \in C[-1, 1]$ and $x \in [-1, 1]$,

$$| f(x) - L_n(f, x) | \le \tilde{\omega}\left(f, | x | (1 - \rho_{1,n}) + \sqrt{1 - x^2}\sqrt{\frac{1 - \rho_{2,n}}{2}}\right),$$

where $\tilde{\omega}(f, t)$ is the lest concave majorant of the first modulus of continuity.

Lehnhoff obtained better estimates. He constructed operators by means of convolution with Matsuoka kernels.

Define

$$H_n(f, x) = \frac{1}{\pi} \int_{-\pi}^{\pi} f(\cos(arccos(x+v))) K_{3n-3}(v) dv \qquad (5.38)$$

where

$$K_{3n-3}(t) = \frac{10}{n(11n^4 + 5n^2 + 4)} \left(\frac{\sin(nt/2)}{\sin(t/2)} \right)^6.$$

It is a special case of the Jackson-Matsuoka kernels (2.8).

Theorem 5.7.4 (Lehnhoff, [230]). *For every function f continuous on $[-1, 1]$ and any natural n, there is an algebraic polynomial $H_n(f)$ of degree $3n-3$ such that, for all $x \in [-1, 1]$,*

$$| f(x) - H_n(x) | \leq 2\omega \left(f, \sqrt{\frac{30}{11}} \frac{|x|}{n^2} + \sqrt{\frac{20}{11}} \frac{\sqrt{1-x^2}}{n} \right)$$

$$\leq 4 \left(\omega \left(f, \frac{|x|}{n^2} \right) + \omega \left(f, \frac{\sqrt{1-x^2}}{n} \right) \right).$$

Now set

$$M_n(f, x) = H_n(f, x) + \frac{1+x}{2} [f(1) - H_n(f, 1)] + \frac{1-x}{2} [f(-1) - H_n(f, -1)].$$

Theorem 5.7.5 (Lehnhoff, [231]). *For $n \geq 10$, $M_n : C[-1, 1] \to \Pi_{3n-3}$ and for each $f \in C[-1, 1]$,*

$$| f(x) - M_n(f, x) | \leq 10 \omega \left(f, \frac{\sqrt{1-x^2}}{n} \right).$$

Theorem 5.7.6 (Gonska, [144]). *If H_n is defined by (5.38), then each $f \in C[-1, 1]$,*

$$| f(x) - H_n(x) | \leq 1.66 \, \tilde{\omega} \left(f, \frac{\sqrt{1-x^2}}{n} + \frac{|x|}{n^2} \right),$$

where $\tilde{\omega}(f, t)$ is the least concave majorant of the modulus of continuity.

Balázs and Kilgore justified that it is important to investigate the constants in the results related with simultaneous approximation and they began to study the problem. They proved a new identity for the derivative of a trigonometric polynomial, based on a well-known identity of M. Riesz, and provided a new proof of Gopengauz's theorem which reduces the problem of estimating the constant there to the question of estimating the constant in Trigub's theorem. The original proofs of these results (and of related works) are uneconomical concerning constants.

Theorem 5.7.7 (Balázs-Kilgore, [10]). *For a function $f \in C^r[-1,1]$ let P_n be a polynomial satisfying (2.31) for some C (which may or may not depend upon n or f, as we choose) and also satisfying*

$$f^{(k)}(\pm 1) = P_n^{(k)}(\pm 1), \qquad 0 \leq k \leq r. \tag{5.39}$$

Then

$$\mid f^{(k)}(x) - P_n^{(k)}(x) \mid \leq K \, (\delta_n(x))^{r-k} \, \omega\left(f^{(r)}, \delta_n(x)\right),$$

with $K \leq \max(4e^e C, 7C + 7)$. In particular, the relation between K and C is absolute and independent of all other quantities involved.

Notice that we can use this result to obtain a new proof of Gopengauz's theorem. Balázs and Kilgore constructed new polynomials satisfying (2.31) and (5.39).

Theorem 5.7.8 (Bashmakova, [16]). *For $f \in C[-1,1]$, there exists a sequence $\{L_n(f)\}$ of linear polynomial operators, $L_n : C[-1,1] \rightarrow \mathbb{P}_n$, such that for $f \in C[-1,1]$ and $x \in [-1,1]$,*

$$\mid f(x) - P_n(f,x) \mid \leq \left(\frac{19}{16} + \frac{A}{(n+1)^{1-3\alpha}}\right)$$

$$\times \omega\left(f, \frac{\pi\sqrt{1-x^2}}{n+1} + \frac{\pi}{(n+1)^{1+\alpha}} + \frac{3\pi^2}{2(n+1)^2}\right),$$

where $0 < \alpha < 1/3$ and A is an absolute constant.

5.8   The boolean sums approach

The boolean sum of two operators A and B is defined by

$$A \oplus L = A + L - A \circ L, \tag{5.40}$$

whenever it makes sense.

For a sequence of operators $\{L_n\}$, $L_n : C[-1,1] \rightarrow C[-1,1]$, and L defined by (5.1), three types of boolean sums can be considered,

$$L \oplus L_n, \quad L_n \oplus L \quad \text{and} \quad L \oplus L_n \oplus L.$$

These are motivated by the following result.

Theorem 5.8.1 (Cao and Gonska, [63]). *Let P and Q be linear operators mapping a function space G (consisting of functions on the domain D) into a subspace H of G. Let G_0 be a subset of G, and let $\mathcal{L} = \{l\}$ be a set of linear functionals defined on H.*

  (i) *Let $l(Pf) = l(f)$ for all $l \in \mathcal{L}$ and all $f \in H$. Then $l((P \oplus Q)f) = l(f)$ for all $l \in \mathcal{L}$ and all $f \in H$.*

(ii) *Let $Qf = f$ for all $f \in G_0$. Then $(P \oplus Q)f = f$ for all $f \in G_0$.*

(iii) *Let f and Qf be in the set of all functions g such that $Pg = g$. Then $(P \oplus Q)f = f$.*

In other words, $P \oplus Q$ inherits certain interpolation properties of P, the function precision of Q, and also some function precision properties of P.

In 1983 Gonska and Hinnemann used the DeVore operators to obtain polynomials to approximate differentiable functions with a better rate.

Theorem 5.8.2 (Gonska and Hinnemann, [159]). *Let $r \geq 0$. For each $n \geq 4(r + 2)$ there exists a linear operator $Q_n : C^r[-1, 1] \to \mathbb{P}_n$ such that*

$$| f(x) - Q_n(f, x) | \leq C_r \, (\delta_n(x))^r \, \omega_2 \left(f^{(r)}, \delta_n(x) \right), \qquad (5.41)$$

for all $f \in C^r[-1, 1]$ and each $x \in [-1, 1]$, where the constant C_r depends only on r.

Later, in 1985, Gonska and Hinnemann showed that generalization to a higher-order modulus is possible, if we consider linear operators defined for differentiable functions. They used a smoothing method. They first approximated the functions by some special differentiable functions. In particular, they considered a theorem of Müller and an easy corollary that follows from the properties of the moduli of smoothness (with the convention $\omega_0(f, t) = \|f\|$).

Theorem 5.8.3 (Müller, [265]). *Given $r \in \mathbb{N}_0$ and $s \in \mathbb{N}$, there exists a constant $C = (r, s)$ such that, for each $h \in (0, 2]$, one has a map $F_h = F_{h,r+s} : C^r[-1, 1] \to C^{2r+s}[-1, 1]$ with the following properties: for all $f \in C^r[-1, 1]$,*

$$\|f^{(i)} - F^{(i)}_{r,r+s}\| \leq C w_{r+s-i}(f^{(i)}, h), \quad 0 \leq i \leq r,$$

and

$$\|F^{(r+s)}_{r,r+s}\| \leq C h^{-(r+s)} w_{r+s}(f, h).$$

Corollary 5.8.4. *Under the conditions of (5.8.3),*

$$\|f^{(i)} - F^{(i)}_{r,r+s}\| \leq C_{r,s} h^{r-i} w_s(f^{(r)}, h), \quad 0 \leq i \leq r,$$

and

$$h^s \|F^{(r+s)}_{r,r+s}\| \leq C_{r,s} \, w_s(f^{(r)}, h),$$

with a different constant.

Let us recall a result of Trigub.

Proposition 5.8.5 (Tribug, [388]). *For each $m, n, p \in \mathbb{N}$, there exists $T_{n,p} \in \mathbb{P}_n$ such that, for $x \in [-\lambda, \lambda]$ $(\lambda > 0)$,*

$$| x^p - x^{p+2m} T_{n,p}(x^2) | < \frac{C_{m,p} \lambda^p}{n^p},$$

where $C_{m,p}$ depends only on m and p.

The construction of Gonska and Hinnemann goes as follows:

a) Set $p = r+s$ and let $\{M_n\}$, $(M_n : C[-1,1] \to \mathbb{P}_n$, $n \geq p-1)$, be any sequence of linear operators satisfying

$$| f(x) - M_n(f,x) | \leq C_p \omega_p(f, \Delta_n(x)),$$

for all $f \in C[-1,1]$ and $x \in [-1,1]$.

b) Let $H : C^r[-1,1] \to \Pi_{2r+1}$ be the Hermite interpolation operator which, for $0 \leq k \leq r$, gives

$$H^{(k)}(f, \pm 1) = f^{(k)}(\pm 1).$$

It is known that there exist constants A_r and B_r and polynomials $A_i, B_i \in \Pi_{2(r-i)+1}$ $(0 \leq i \leq r)$ such that (see [371]),

$$H(f,x) = \sum_{i=0}^{r}(1 - x^2)^i \left\{ f^{(i)}(1)A_i(x) + f^{(i)}(-1)B_i(x) \right\},$$

where $\|A_i\| \leq A_r$ and $\|B_i\| \leq B_r$.

c) For $n \geq 4(r+1)$ and $0 \leq i \leq r$, let $T_{n,i}$ be the polynomial of Proposition 5.8.5 with $\lambda = 1$, $p = 1$, $m_i = r + 1 - [i/2]$ and $n_i = [n/(4(r+1))]$. Define $R_{n,2} : C^r[-1,1] \to \mathbb{P}_n$ by

$$R_{n,2}(f,x) = \sum_{i=0}^{r}(1 - x^2)^{r+1+i-[i/2]}T_{n,i}(x)\{f^{(i)}(1)A_i(x) + f^{(i)}(-1)B_i(x)\},$$

where A_i and B_i are given in b).

It can be proved that $R_{n,2} \in \mathbb{P}_n$, for $n \geq 4(r+1)$ and $R_{n,2}^{(k)}(f, \pm 1) = 0$, $0 \leq k \leq r$.

d) Define $R_n : C^r[-1,1] \to \mathbb{P}_n$ as $R_n = H - R_{n,2}$ and let Q_n be the boolean sum of R_n and M_n (see (5.40)).

For each $f \in C^r[-1,1]$ and $0 \leq k \leq r$, one has $Q_n^{(k)}(f, \pm 1) = f^{(k)}(f, \pm 1)$.

Theorem 5.8.6 (Gonska and Hinnemann, [147]). *Assume that $r \geq 0$ and $s \geq 1$ and let the sequence of linear operators $\{Q_n\}$ be defined by (5.40).*

(i) *There exists a constant $M_{r,s}$ such that, for $n \geq \max\{(4(r+1), r+s\}$, $0 \leq k \leq r$, $f \in C^r[-1,1]$ and $x \in [-1,1]$ one has*

$$| f^{(k)}(x) - Q_n^{(k)}(f,x) | \leq M_{r,s}(\Delta_n(x))^{r-k}\,\omega_s(f^{(r)}, \Delta_n(x)).$$

(ii) *If $r \geq s \geq 1$ and $n \geq 4(r+1)$, there exists a constant $M_{r,s}$ such that for $f \in C^r[-1,1]$, $0 \leq k \leq r - s$ and $x \in [-1,1]$ one has*

$$| f^{(k)}(x) - Q_n^{(k)}(f,x) | \leq M_{r,s}(\delta_n(x))^{r-k}\,\omega_s(f^{(r)}, \delta_n(x)).$$

Define

$$G_{m(n)}(f,x) = \frac{1}{\pi} \int_{-\pi}^{\pi} f(\cos(\arccos(x + v)))\,K_{m(n)}(v)dv, \qquad (5.42)$$

where $K_{m(n)}$ is the Matsuoka kernel (2.8).

In 1986, Cao and Gonska began to publish a series of paper devoted to study of the boolean sums of positive linear operators. In the first paper they gave an upper bound for the local degree of approximation by the boolean sum of positive linear operators in terms of the second-order modulus of continuity of the function [62]. In particular, they applied the main result to study the Pichugov-Lehnhoff operators presented above. Other results were given by Gonska in [145].

Define

$$G_{m(n)}(f,x) = \frac{1}{\pi} \int_{-\pi}^{\pi} f(\cos(\arccos(x+v))) K_{m(n)}(v) dv, \qquad (5.43)$$

where $K_{m(n)}$ is the Matsuoka kernel (2.8).

By applying Theorem 5.8.1 to $G_{m(n)}$ one obtains the following corollary. We use the notation

$$G_{m(n)}^+ = L \oplus G_{m(n)} \quad \text{and} \quad G_{m(n)}^1 = L \oplus G_{m(n)} \oplus L.$$

Corollary 5.8.7. *The operator $G_{m(n)}^+$ has the following properties:*

(i) $G_{m(n)}^+(f,\pm1) = f(\pm1)$, *for all $f \in C[-1,1]$.*

(ii) $G_{m(n)}^+ f = f$ *for all $f \in \mathbb{P}_1$.*

(iii) $G_{m(n)}^+ = G_{m(n)}^1$.

Theorem 5.8.8 (Cao and Gonska, [63]). *Let $n \geq 2$, $m(n) \in \mathbb{N}$, and $C_1 n \leq m(n) \leq C_2 n$. Furthermore, let $A_n : C[-1,1] \to \mathbb{P}_n$ be a sequence of positive linear operators, satisfying the conditions*

(i) $A_n(1,x) = 1$,

(ii) $A_n(t,x) = \lambda_n x$, *where $1 - \lambda_n = \mathcal{O}(1/n^2)$,*

(iii) $A_n((t-x)^2, x) = \mathcal{O}((1-x^2)/n^2 + 1/n^4)$.

Then we have for all $f \in C[-1,1]$ and all $x \in [-1,1]$ that

$$| f(x) - A_n^+(f,x) | \leq C\omega_2\left(f, \frac{\sqrt{1-x^2}}{n}\right).$$

From this one obtains

Theorem 5.8.9 ([63]). *Assume $n \geq 2$ and $s \geq 3$. If G_{ns-s} is defined by (5.42), then*

$$| f(x) - G_{ns-s}(f,x) | \leq C\omega_2\left(f, \frac{\sqrt{1-x^2}}{n}\right)$$

for all $f \in C[-1,1]$ and $x \in [-1,1]$.

Similar estimates hold for the corresponding operators $G_{m(n)}^1$.

Some extensions of Theorem 5.8.8 appeared in [65].

In [66] Cao and Gonska realized another construction by considering again the Jackson-Matsuoka kernels, but of a higher order. In both cases the use of methods from Fourier Analysis and standard ideas of Numerical Analysis was very important. One of the results of [64] was generalized in [67] as follows.

Theorem 5.8.10. *Let $A_n : C[a,b] \to C^1[a,b]$ be a sequence of positive linear operators satisfying the following conditions:*

  (i) $A_n(1,x) = 1$, $x \in [a,b]$.
  (ii) *For $x \in [a,b]$ and $0 \le \varepsilon_n \le 2$,*

$$A_n(|\,t - x\,|, x) \le C(\varepsilon_n \sqrt{(x-a)(b-x)} + \varepsilon_n^2).$$

(iii) *For all $h \in C^1[a,b]$,*
$$\|dA_n(h,x)/dx\| \le C\|h'\|.$$

Then, for all $f \in C[a,b]$,

$$|\,f(x) - A_n^*(f,x)\,| \le C\omega\left(f, \varepsilon_n \sqrt{(x-a)(b-x)}\right).$$

In 1990, Cao and Gonska looked for general conditions in order to find boolean sums of linear operators that satisfy Teliakovskii-type estimates. The main result asserts that, if A_n is a sequence of polynomial linear operators for which a Timan-type estimate holds, then one can always derive a Teliakovskii-type estimate for their boolean sum modification A_n^+. This result can be applied to some of the operators presented in a previous section to obtain operators with a Teliakovskii-type estimate.

Theorem 5.8.11 (Cao and Gonska, [67]). *For each $n \in \mathbb{N}$ fix $m(n) \in \mathbb{N}_0$ such that $Cn \le m(n) \le C_2 n$ for some positive constants C_1 and C_2. Let $A_n : C[-1,1] \to \mathbb{P}_{m(n)}$ satisfying the Timan estimate*

$$|\,f(x) - A_n(f,x)\,| \le C_3\omega\left(f, \frac{\sqrt{1-x^2}}{n} + \frac{1}{n^2}\right),$$

for $f \in C[-1,1]$ and $x \in [-1,1]$. Then there exists a constant C_4 such that, for all $f \in C[-1,1]$ and $x \in [-1,1]$,

$$|\,f(x) - A_n^+(f,x)\,| \le C_4\omega\left(f, \frac{\sqrt{1-x^2}}{n}\right).$$

In Theorem 5.8.9 Jackson-Matsuoka kernels were studied. Boos, Cao and Gonska extended the result to the case when in the boolean sums we consider convolution with arbitrary positive kernels.

Theorem 5.8.12 (Boos, Cao and Gonska, [35]). *Let $m(n) \geq 2$, let the even kernels $K_{m(n)}$ in (5.49) satisfy $K_{m(n)}(v) \geq 0$ and*

(i) $$\sqrt{1 - \rho_{1,\,m(n)}} \leq C_1 \alpha_n,$$

(ii) $$\sqrt{1 - \rho_{2,\,m(n)}} \leq C_2 \alpha_n,$$

(iii) $$\frac{3}{2} - 2\rho_{1,\,m(n)} + \frac{1}{2}\rho_{2,\,m(n)} \leq (C_3 \beta_n)^4,$$

where $0 < \tau_n = \max\{\alpha_n, \beta_n\} \leq 1$. Let $G_{m(n)}(f, t)$ be defined by

$$G_{m(n)}(f, t) = \frac{1}{\pi} \int_{-\pi}^{\pi} f(\cos(\arccos(x + t)))K_{(n-2)p}(t)dt. \tag{5.44}$$

Then for $f \in C[-1, 1]$, $n \in \mathbb{N}$ and $x \in [-1, 1]$, one has

$$| G_{m(n)}^+(f, x) - f(x) | \leq M \omega_2(f, \tau_n \sqrt{1 - x^2}),$$

where the constant M is determined by

$$M = 3 + \frac{3}{2} \max\left\{C_1^2 + \frac{1}{4}C_2^2 + \frac{3}{2}C_3^4,\ 2C_1^2 + \sqrt{2}C_2 + \frac{1}{2}C_2^2 + \frac{1}{2}C_3^4\right\}.$$

If we do not assume (ii), then a similar inequality holds with a bigger constant.

The authors used the last theorem to give explicit values of the constant C in Theorem 5.8.9. For instance, for the Jackson-Matsuoka kernel K_{3n-3}, one has $C < 15$. They also studied the asymptotic of the constants.

There are several interesting consequences.

Corollary 5.8.13. *Assume $m(n) \geq 2$ and $K_{m(n)} \geq 0$. Let $\{\varepsilon_n\}$ $(0 < \varepsilon \leq 1)$ be a sequence such that*

(i) $1 - \rho_{1,m(n)} = \mathcal{O}(\varepsilon_n^2)$,

(ii) $\frac{3}{2} - 2\rho_{1,m(n)} + \frac{1}{2}\rho_{2,m(n)} = \mathcal{O}(\varepsilon_n^4)$.

Then for $f \in C[-1, 1]$, $n \in \mathbb{N}$ and $x \in [-1, 1]$, one has

$$| G_{m(n)}^+(f, x) - f(x) | \leq M \omega_2(f, \varepsilon_n \sqrt{1 - x^2}).$$

Cao and Gonska also investigated Fejér-Korovkin kernels (of higher order) in detail where, for $p \in \mathbb{N}$ fixed and ≥ 2, these kernels are given by

$$D_{(n-a)p}(v) = \left(\frac{\cos(nv/2)}{n^2(\cos v - \cos(\pi/n))}\right)^{2p}$$

and these are the pth powers of the ordinary Fejér-Korovkin kernels (apart from constants). Let $K_{(n-2)p}$ be a normalization of $D_{(n-a)p}$ such that

$$\frac{1}{\pi} \int_{-\pi}^{\pi} K_{(n-a)p}(t)dt = 1.$$

With this notation define the operator

$$F_{(n-2)p}(f,x) = \frac{1}{\pi} \int_{-\pi}^{\pi} f(\cos(\arccos(x+t)))K_{(n-2)p}(t)dt.$$

The next result gives an estimate for the boolean sums of these operators. The original paper includes an analysis of the asymptotic of the constant.

Theorem 5.8.14 (Cao, Gonska and Wenz, [73]). *Let $n \geq 3$ and $p \geq 2$. Then for $f \in C[-1,1]$ there holds*

$$| F_{(n-2)p}^+(f,x) - f(x) | \leq C_p\, w_2(f, \sqrt{1-x^2}/n).$$

In 1996, Cao and Gonska gave several results where the constants in a Teliakovskii-type estimate is taking into account. They also studied the asymptotic of the constants. Such results will not be included here.

Theorem 5.8.15 (Cao and Gonska, [71]). *For each $n \in \mathbb{N}$, let $K_{m(n)} \geq 0$ and $G_{m(n)}$ be given as in (5.45) and (5.45) respectively. Then for $f \in C[-1,1]$, $x \in [-1,1]$ and $h > 0$, one has*

$$| f(x) - G_{m(n)}^+(f,x) | \leq \left[2 + \left(2 + 2\sqrt{2}\right)\frac{\sqrt{1-\rho_{1,m(n)}}}{h}\right] w(f, h\sqrt{1-x^2}).$$

If $\rho_{1,m(n)} \geq 0$, then the constant can be taken as

$$2 + \left(3 + \sqrt{2}\right)\frac{\sqrt{1-\rho_{1,m(n)}}}{h}.$$

If we consider Fejér-Korovkin kernels of the form

$$K_n(v) = \frac{1}{n+1}\left(\frac{\sin(\pi/(n+2))\cos((n+2)v/2)}{\cos v - \cos(\pi/(n+2))}\right)^2$$

and W_n^* is the corresponding boolean sum, then

$$| f(x) - W_n^+(f,x) | \leq 12\, w\left(f, \frac{\sqrt{1-x^2}}{n+2}\right)$$

and

$$| f(x) - W_n^+(f,x) | \leq 6\, w\left(f, \frac{\pi\sqrt{1-x^2}}{n+1}\right).$$

5.9 Discrete operators

As Vértesi [398] showed, we cannot have linear operators with $n + 1$ equidistant nodes satisfying DeVore-Gopengauz inequalities (see also [139]).

Discrete versions of operators $G_{m(n)}$ and $G_{m(n)}^r$ were investigated in [64], [66] and [68]. They considered special positive algebraic convolution integrals and constructed a discrete version by using appropriated numerical quadrature.

Let us state the following problem. Can we find a triangular matrix of distinct nodes $\{x_{k,n}\}$ ($k = 0, \ldots, n$, $-1 \leq x_{k,n} \leq 1$), and a triangular matrix of positive functions $\{\varphi_{k,n}\}$ ($k = 0, \ldots, n$, $n \in \mathbb{N}$) defined on $[-1, 1]$ such that, for all $f \in C[-1, 1]$ satisfying $\omega_2(f, t) \leq Ct^\alpha$ ($0 < \alpha \leq 2$) one has

$$\|f - L_n(f)\| = \mathcal{O}(n^{-\alpha}),$$

where

$$L_n(f, x) = \sum_{k=0}^{n} f(x_{k,n}) \varphi(x)?$$

This problem was stated at the end of a paper by Butzer, Stens and Wehrens in 1979 [58]. They asked for a constructive proof and remarked that Bernstein operators

$$B_n(f, x) = \sum_{k=0}^{n} \binom{n}{k} \left(\frac{k}{n}\right) x^k (1 - x)^{n-k}$$

do not provide a solution, since these only give the rate $\mathcal{O}(n^{-\alpha/2})$.

Versions of this question were raised before by other authors.

Notice that we can not use boolean sums of positive operators that are not positive linear operators. Gonska and Zhou [149] formulated another question:

Do there exist positive linear operators $L_n : C[-1, 1] \to \mathbb{P}_n$ such that, for all $f \in C[-1, 1]$ and $x \in [-1, 1]$, one has

$$| f(x) - L_n(f, x) | \leq C \omega_2 \left(f, \frac{\sqrt{1 - x^2}}{n}\right),$$

where the constant is independent of f, n and x?

We can also ask for a solution of the last problem with discretely defined operators. It is called the *strong form of Butzer's problem*.

In 1981, Butzer and Wehrens provided a theoretical solution [61]. They constructed a sequence $\{L_n\}$ by applying the Christoffel quadrature formula to positive polynomial convolution integrals in the Legendre transform setting. They also stated another problem: is it possible to construct a sequence of positive linear operators $U_n : C[-1, 1] \to \mathbb{P}_n$ such that there exist non-constant functions $f \in C[-1, 1]$ for which $\|f - U_n(f)\| = \mathcal{O}(n^{-2})$? Since the nodes can not be calculated exactly and the coefficient of the fundamental polynomials are not known, their solution is not a constructive one.

In [64] Cao and Gonska introduced certain sequences of discrete positive linear operators. Taking into account the drawbacks in the Butzer and Wehrens solution, they looked for special discrete versions of convolution-type operators. In particular they used the Jackson-Matsuoka kernels (2.8).

For $s = 3$, Matsuoka [250] found an exact expression for the coefficients $\rho_{k,3n-3}$ of the kernel K_{3n-3} in the expansion

$$K_{3n-3}(t) = \frac{1}{2} + \sum_{k=1}^{3n-3} \rho_{k,3n-3} \cos(kt), \qquad n \geq 1.$$

Thus we can use the fundamental polynomials

$$A_{r,\,3n-3,\,N_0}(x) = \frac{1}{N_0} \left(1 + \sum_{k=1}^{3n-3} \rho_{k,3n-3}\, T_k(x_{r,N_0}) T_k(x) \right)$$

for $1 \leq r \leq N_0$, where

$$x_{r,N_0} = \cos \frac{2r-1}{2N_0}\pi, \quad 1 \leq r \leq N_0,$$

and T_k is the Chebyshev polynomial.

The parameter N_0 appeared because of the Gaussian quadrature to be used in the discrete version of the corresponding convolution operator.

Now, if we define

$$\Lambda_{3n-3,\,N_0}(f,x) = \sum_{r=1}^{N_0} f(x_{r,N_0}) A_{r,\,3n-3,\,N_0}(x),$$

we obtain a positive linear operator. For these operators and some boolean sums, modification of them by Cao and Gonska proved pointwise Jackson-type theorems of Gonpengauz type involving the first- and second-order moduli of smoothness. In particular, the following result was given:

Theorem 5.9.1 (Cao and Gonska, [64]). *If $N_0 \geq 3n/2$, $0 < \alpha \leq 2$, $f \in C[-1,1]$ and $\omega_2(f,t) \leq Ct^\alpha$, then*

$$\|f - \Lambda_{3n-3,\,N_0}(f)\| \leq Cn^{-\alpha}.$$

In [68] Cao and Gonska considered more general kernels.

For $m \in \mathbb{N}$, let

$$K_m(t) = \frac{1}{2} + \sum_{k=1}^{m} \rho_{k,m} \cos(kt) \tag{5.45}$$

be an even positive trigonometric polynomial of degree m. Define a polynomial operator $G_m : [-1,1] \to \mathbb{P}_m$ by

$$G_m(f,x) = c_0(f) + \sum_{k=0}^{m} \rho_{k,m}\, c_k(f) T_k(x), \qquad f \in C[-1,1], \tag{5.46}$$

where $c_k(f)$ is the kth coefficient of f in its Chebyshev-Fourier expansion (see (3.10)) and T_k is the Chebyshev polynomial. G_m is a positive linear operator which reproduces constant functions. If the coefficients $c_k(f)$ are known, then $G_m(f)$ can be efficiently computed using Clensaw's algorithm (see [409]).

The operator G_m is not a discrete one. But some numerical formulas can be used to discretize (3.10). Let Q_N a numerical quadrature of the form

$$Q_N(g) = \sum_{j=0}^{N+1} \beta_{j,N} g(x_{j,N}) \tag{5.47}$$

with nodes $-1 \le x_{0,N} < x_{1,N} < \cdots < x_{N+1,N} = 1$ and apply it to

$$\int_{-1}^{1} \frac{g(u)du}{\sqrt{1-u^2}}.$$

Thus, we write

$$\int_{-1}^{1} \frac{g(u)du}{\sqrt{1-u^2}} = Q_N(g) + R_N(g),$$

where $R_N(g)$ is the error. We assume that Q_N is of exact degree $d(Q_N)$. That is $R_N(p) = 0$ for each polynomial $P \in \mathbb{P}_{d(Q_N)}$ and there exists a polynomial $q \in \mathbb{P}_{d(Q_N)+1}$ such that $R_N(q) \ne 0$.

With the notation given above we define an operator $\Lambda[K_m, Q_N]$ as follows: for each $f \in C[-1,1]$ and $x \in [-1,1]$,

$$\Lambda[K_m, Q_N](f, x) = \frac{1}{\pi} Q_N(f) + \frac{2}{\pi} \sum_{k=1}^{m} \rho_{k,m} \, Q_N(f \, T_k) \, T_k(x). \tag{5.48}$$

Theorem 5.9.2 (Cao and Gonska, [68]). *Let K_M be a positive kernel with $\rho_{1,M} \ge 0$ and let Q_N (degree$(Q_N) \ge M + 2$) and $\Lambda[K_M, Q_N]$ be given by (5.47) and (5.48) respectively. Then for all $f \in C[-1,1]$ one has*

$$\|f - \Lambda[K_M, Q_N](f)\| \le 5\omega_2(f, \sqrt{1-\rho_{1,M}}) + 2\sqrt{1-\rho_{1,M}} \, \omega_1(f, \sqrt{1-\rho_{1,M}}).$$

Theorem 5.9.3 ([68]). *Let $K_{m(n)}$ be a sequence of positive kernels satisfying $1 - \rho_{1,m(n)} = \mathcal{O}(n^{-2})$. Suppose, furthermore, that $\{Q_N\}$ is an associated sequence of positive quadrature sums satisfying (degree$(Q_N) \ge m(n) + 2$). Then for all $f \in C[-1,1]$ for which $\omega_2(f,t) \le Ct^\alpha$, $0 < \alpha \le 2$, one has*

$$\|f - \Lambda[K_{m(n)}, Q_N](f)\| \le C \, n^{-\alpha}.$$

With some additional assumptions on $K_{m(n)}$ we can obtain pointwise improvements at the endpoints.

Theorem 5.9.4 ([68]). *Let $K_{m(n)}$ be a sequence of positive kernels with $m(n) \geq 2$, for $n \in \mathbb{N}$ and let $\{Q_N\}$ be an associated sequence of positive quadrature sums satisfying $(degree(Q_N) \geq m(n) + 2)$. Furthermore, suppose that*

$$1 - \rho_{1,m(n)} = \mathcal{O}(n^{-2})$$

and

$$\frac{3}{2} - 2\rho_{1,m(n)} + \frac{1}{2}\rho_{2,m(n)} = \mathcal{O}(n^{-4}).$$

Then for all $f \in C[-1,1]$, $n \geq 2$ and $x \in [-1,1]$, one has

$$| f - \Lambda[K_{m(n)}, Q_N](f,x) |$$

$$\leq C\left(\omega_2\left(f, \frac{\sqrt{1-x^2}}{n} + \frac{|x|}{n} \right) + \frac{|x|}{n\sqrt{1-x^2} + |x|}\omega_1\left(f, \frac{\sqrt{1-x^2}}{n} + \frac{|x|}{n} \right) \right).$$

There are several examples of kernels for which the conditions assumed above hold. We present some of them.

(1) **B-Z kernels:** The Bohman and Zheng Wei-xing kernel is defined by [346]

$$Z_n(x) = \left(\frac{\cos((n+1)x/2)}{\cos x - \cos(\pi/(n+1))} \right)^2$$

$$\times \left(\frac{n+1}{\pi} \sin\frac{\pi}{n+1}\left(1 - \frac{\pi}{n+1}\cot\frac{\pi}{n+1} \right) + \left(1 - \frac{n+1}{2\pi}\sin\frac{2\pi}{n+1} \right) \right).$$

In this case

$$\rho_{k,n} = \left(1 - \frac{k}{n+1} \right)\cos\frac{k\pi}{n+1} + \frac{1}{\pi}\sin\frac{k\pi}{n+1}, \qquad 1 \leq k \leq n.$$

(2) **General Korovkin kernels:** Fix $\Phi \in C[0,1]$ such that, for each $n \in \mathbb{N}$,

$$c_n = \sum_{k=0}^{n} \Phi^2(k/n) > 0.$$

Define

$$K_n(t) = \frac{1}{2c_n}\left| \sum_{k=0}^{n} \Phi(k/n)e^{ikt} \right|^2.$$

Since K_n is a non-negative trigonometric polynomial of degree not greater than n, it can be written in the form

$$K_n(t) = \frac{1}{2} + \sum_{k=1}^{n} \rho_{k,n}\cos(kt). \tag{5.49}$$

If $\phi \in \text{Lip}_1[0,1]$ and $\int_0^1 \Phi^2(t)dt > 0$, then

$$n^1(1 - \rho_{1,n}) \le C \left(\int_0^1 \Phi^2(t)dt \right)^{-1}.$$

For $\Phi(t) = \sin(\pi t)$, we obtain the Fejér-Korovkin kernel F_{n-2}. The following result is announced in [73].

Theorem 5.9.5. *If the operators G_n^+ based upon the Fejér-Korovkin kernels $K_n(v)$ are denoted by F_n^+, then one has*

$$\mid f(x) - F_n^+(f, x) \mid \le 12\,\omega_1\left(f, \sqrt{1 - x^2}/n\right), \tag{5.50}$$

for all $f \in C[-1, 1]$, all $\mid x \mid \le 1$, and all $n \in \mathbb{N}$.

(3) **Jackson-Matsuoka kernels:** These were presented before. In this case $1 - \rho_{1,n} = \mathcal{O}(n^{-2})$.

(4) **Jackson-de la Vallée-Poussin kernels:** These are defined by

$$P_{2n-1}(x) = \frac{2 + \cos x}{4n^3} \left(\frac{\sin((nx)/2)}{\sin(x/2)} \right)^4.$$

It can be proved that $1 - \rho_{1,2n-1} \le 3/(2n^2)$.

Notice that in the results presented above the estimates are given in terms of $1 - \rho_{1,m(n)}$. The operators do not interpolate at the endpoints 1 and -1, thus Teliakovskii-type estimates can not be obtained with such a construction. This explains one of the reasons for Cao and Gonska to use the boolean sums approach.

In 1995, Cao and Gonska noticed that the discrete version given by

$$\Lambda_{m(n),\,N_0}(f, x) = \frac{1}{N_0} \sum_{r=1}^{N_0} f(x_{r,N_0}) \left\{ 1 + 2 \sum_{k=1}^{m(n)} \rho_{k,\,m(n)} T_k(x_{r,N_0}) T_k(x) \right\} \tag{5.51}$$

is equivalent to the operator $G_{m(n)}$. They also studied the saturation order.

Theorem 5.9.6 (Cao and Gonska, [70]). *If $N_0 \ge m(n) + 1$, $\int_0^\pi \mid K_{m(n)}(t) \mid dt = \mathcal{O}(1)$ and $\Lambda_{m(n),\,N_0}$ is defined by (5.51), then there exist positive constants C_1 and C_2 such that, for all $f \in C[-1, 1]$,*

$$C_1\|f - G_{m(n)}(f)\| \le \|f - \Lambda_{m(n),\,N_0}(f)\| \le C_2\|f - G_{m(n)}(f)\|.$$

Cao and Gonska [70] also noticed that many solutions to Butzer's problem can be obtained from the periodical case.

Gavrea [135] was the first to construct positive linear operators which yield the interpolatory estimates

$$\mid f(x) - P_n(x) \mid \le C\omega\left(f, \frac{\sqrt{1 - x^2}}{n}\right). \tag{5.52}$$

In [136], he also constructed operators which yield

$$| f(x) - P_n(x) | \leq C\omega_2 \left(f, \frac{\sqrt{1 - x^2}}{n} \right). \tag{5.53}$$

Let $L_m : C[0, 1] \to \mathbb{P}_m$ be defined by

$$L_m(f, x) = f(0)(1 - x)^m + f(1)x^m$$
$$+ (m - 1) \sum_{k=1}^{m-1} p_{m,k}(x) \int_0^1 p_{m-2,k-1}(t)f(t)dt, \tag{5.54}$$

where

$$p_{m,k} = \binom{m}{k} x^k (1 - x)^{m-k}. \tag{5.55}$$

Fix a polynomial $P_m \in \mathbb{P}_m$, $P_m(x) = \sum_{k=0}^m a_{m,k} x^k$, such that

$$P_m(x) \geq 0, \quad \text{for all} \quad x \in [0, 1] \qquad \int_0^1 P_m(x)dx = 1$$

and

$$P_m'(x) \geq 0, \quad \text{for all} \quad x \in [0, 1].$$

Now, define an operator $H_{m+2} : C[0, 1] \to \mathbb{P}_{m+2}$ by

$$H_{m+2}(f, x) = \sum_{k=0}^m \frac{a_{m,k}}{k + 1} L_{k+2}(f, x). \tag{5.56}$$

Theorem 5.9.7 (Gavrea, [136]). *Let the operators H_{m+2} be defined as in (5.56), then for each $f \in C[0, 1]$ and $x \in [0, 1]$ one has*

$$| f(x) - H_{m+2}(f, x) | \leq \frac{9}{4}\omega_2 \left(f, \sqrt{x(1 - x)}\sqrt{1 - \int_0^1 t^2 p_m(t)dt} \right).$$

In [137] Gavrea, Gonska and Kacsó constructed some operators $T_{2n+1} : C[0, 1] \to \mathbb{P}_{2n+1}$ of the form

$$T_{2n+1}(f, x) = \sum_{k=0}^n q_{2n+1,k}(x)f\left(\frac{k}{n} \right),$$

for which the inequality

$$| f(x) - T_{2n+1}(f, x) | \leq C\omega_2 \left(f, \frac{\sqrt{\alpha_n(x)}}{n} + \frac{\sqrt{1 - x^2}}{n} \right)$$

holds, where $\alpha_n(x)$ is a bounded function such that $\alpha_n(0) = \alpha_n(1) = 0$.

Recall that Gavrea constructed in 1996 non-discrete positive linear operators satisfying DeVore-Gopengauz inequalities in terms of the second-order modulus of continuity. In 1998 Gavrea, Gonska and Kacsó presented, for the first time, positive linear operators with equidistant nodes solving Butzer's problem in its original form.

For each n, define an operator $S_n : C[0,1] \to C[0,1]$ by

$$S_n(f,x) = \frac{1}{n} \sum_{k=0}^{n} \left[\frac{k-1}{n}, \frac{k}{n}, \frac{k+1}{n}; |t-x|_t \right]_t f\left(\frac{k}{n}\right), \tag{5.57}$$

where, for mutually distinct a, b, c, $[a, b, c[f(t,x)]_t$ means that the divided difference is applied on the variable t.

The next theorem gives sufficient conditions for obtaining operators which solve Butzer's problem.

Theorem 5.9.8 (Gavrea, Gonska and Kacsó, [140]). *Let $L_{m(n)} : C[0,1] \to \mathbb{P}_{m(n)}$ be a sequence of positive linear operators satisfying*

(i) $$L_{m(n)}(1,x) = 1,$$

(ii) $$| L_{m(n)}(t-x,x) | \le C/n^2,$$

(iii) $$| L_{m(n)}((t-x)^2, x) | \le C/n^2,$$

where the constant C is independent of n and x.

Then the operator $\mathcal{L}_{m(n)} = L_{m(n)} \circ S_n$, where S_n is defined by (5.57), satisfies

$$\|f - \mathcal{L}_{m(n)}(f)\| = \mathcal{O}(n^{-\alpha}),$$

for every f for which $\omega_2(f,t) \le Ct^\alpha$ with $0 < \alpha \le 2$.

The last theorem was improved in [138] where characterizations of the solutions of Butzer's problem were given.

As an example, the authors constructed a sequence as follows. First, fix $\lambda \in [-1/2, 1/2]$ and, for each n, fix a polynomial $Q_n(x) = \sum_{k=0}^{n} a_{k,n} x^k$, $a_{n,n} \ne 0$, satisfying the conditions

$$Q_n(x) > 0, \quad \text{for all} \quad x \in [0,1] \quad \text{and} \quad \int_0^1 Q_n(x) x^\lambda (1-x)^\lambda dx = 1.$$

For $f \in C[0,1]$ define

$$L_n^{<\lambda>}(f,x) = \sum_{k=0}^{n} \frac{(\lambda+1)_k}{(2\lambda+2)_k} a_{k,n} D_k^{<\lambda>}(f,x),$$

where $(\lambda)_k = \lambda(\lambda+1)\cdots(\lambda+k-1)$, $(\lambda)_0 = 1$ and

$$D_k^{<\lambda>}(f,x) = \sum_{k=0}^{n} p_{k,n}(x) \frac{\int_0^1 t^\lambda (1-t)^\lambda p_{k,n}(t) f(t) dt}{\int_0^1 t^\lambda (1-t)^\lambda p_{k,n}(t) dt}.$$

Here $p_{k,n}$ is defined by (5.55). These operators were constructed by Lupaş and Mache (see [244], p. 216).

Theorem 5.9.9. *The operators $L_n^{<\lambda>}$ defined above have the following properties:*

(i) *For each n, $L_n^{<\lambda>}$ is positive and $L_n^{<\lambda>} : C[0,1] \to \mathbb{P}_n$.*

(ii) *For $0 < \alpha \le 2$ and every $f \in C[0,1]$ for which $\omega_2(f,t) \le Ct^\alpha$, one has*

$$\|f - (L_n^{<\lambda>} \circ S_n)(f)\| = \mathcal{O}(n^{-\alpha}).$$

Gavrea, Gonska and Kacsó also showed that Theorem 5.9.8 can be applied to the operators constructed by Cao and Gonska. They also construct discretely defined positive linear operators satisfying DeVore-Gopengauz inequalities, generalizing the solution of a strong form of Butzer's problem given earlier by the same authors. They modified some of the ideas used by Gavrea in obtaining Theorem 5.9.7.

Set

$$Q_n^*(x) = \lambda_n^* \, x^d \left(\frac{J_r^{(s,d)}(x)}{x - x_r} \right)^2,$$

where $J_r^{(s,d)}$ is the Jacobi polynomial relative to interval $[0,1]$, x_r is the largest root of $J_r^{(s,d)}$ and λ_n^* is chosen from the condition

$$\frac{1}{s!} \int_0^1 (1-x)^s Q_n^*(x)dx = 1.$$

Define a new polynomial by

$$P_{n+s}^*(x) = \int_0^x \int_0^{t_1} \cdots \int_0^{t_{s-1}} Q_n^*(t_s)dt_s \ldots dt_1 = \sum_{k=0}^{n+s} a_k x^k.$$

The coefficients a_k are used to define the operators

$$H_{n+s+2}^*(f,x) = \sum_{k=0}^{n+s} \frac{a_k}{k+1} L_{k+2}(f,x)$$

$$= (1-x)^2 f(0) \int_0^1 P_{n+s}^*(t(1-x))dt + x^2 f(1) \int_0^1 P_{n+s}^*(tx)dt$$

$$+ \int_0^1 K_{n+s}(x,t)f(t)dt \tag{5.58}$$

where L_k is given by (5.54) and

$$K_{n+s}(x,t) = \sum_{k=0}^{n+s} \sum_{i=1}^{k+1} a_k p_{k+2,i}(x) p_{k,i-1}(t).$$

Consider a quadrature formula

$$\int_0^1 f(x)dx = \sum_{k=1}^{n+s} A_k f(x_k) + R(f) \tag{5.59}$$

with a degree of exactness less than $n + s + 2$. This formula is used to obtain a discrete version of the operators (5.58) by defining

$$\mathcal{H}_{n+s+2}^*(f, x) = (1 - x)^2 f(0) \int_0^1 P_{n+s}^*(t(1 - x))dt \tag{5.60}$$

$$+ x^2 f(1) \int_0^1 P_{n+s}^*(xt)dt + \sum_{k=1}^{n+s} A_k K_{n+s}(x, x_k) f(x_k),$$

where A_k and x_k are the coefficients and the nodes of the quadrature formula (5.59) respectively.

Theorem 5.9.10 (Gavrea, Gonska and Kacsó, [140]). *Let the operators H_{n+s+2}^* and $\mathcal{H}_{n+s+2}^*$ be given by (5.58) and (5.60) respectively. There exists a constant C such that, for every $f \in C[0, 1]$ and $x \in [0, 1]$,*

$$| f(x) - H_{n+s+2}^*(f, x) | \leq C \omega_2 \left(f, \frac{\sqrt{x(1 - x)}}{n} \right)$$

and

$$| f(x) - \mathcal{H}_{n+s+2}^*(f, x) | \leq C \omega_2 \left(f, \frac{\sqrt{x(1 - x)}}{n} \right).$$

They also investigate the potential of these new operators for simultaneous approximation of the first two derivatives. In [72] some other results concerning simultaneous approximation are discussed.

Kacsó obtained discrete operators with the same degree of approximation as Cao and Gonska (in particular, DeVore-Gopengauz inequalities) by using other methods. Moreover, the change of method allows her to present operators which inherit some properties from the initial operators. We will not present here results related with shape preserving approximation.

Let $\Delta_n = \{x_0, x_1, \ldots, x_n\}$ ($x_0 = -1$, $x_n = 1$) be a partition of $[-1, 1]$. For each function $f : [-1, 1] \rightarrow \mathbb{R}$, there exists a unique continuous function $S_{\Delta_n} f$ whose restriction to each one of the intervals $[x_i, x_{i+1}]$ ($0 \leq i \leq n - 1$) is a polynomial of degree not greater than 1 and which interpolates f at the nodes x_i, that is

$$S_{\Delta_n}(f, x_i) = f(x_i), \qquad 0 \leq i \leq n.$$

We use the operator S_{Δ_n} to discretize the operator $G_{m(n)}$. In particular, define

$$\mathcal{G}_{m(n)} = G_{m(n)} \circ S_{\Delta_n}$$

and
$$\mathcal{G}^*_{m(n)} = (L \oplus G_{m(n)}) \circ S_{\Delta_n} = L \oplus \mathcal{G}_{m(n)},$$
where L is given by (5.1).

Theorem 5.9.11 (Kacsó, [186]). *Let the partition Δ_n be given by the points $x_k = \cos\theta_k$, where the point $\theta_k \in [0, \pi]$ satisfies the conditions*

(i) $\theta_k - \theta_{k+1} \leq K/n$, $0 \leq k \leq n-1$,

(ii) $\theta_k/\theta_{k+1} \leq \beta$, $0 \leq k \leq n-2$, *where K and β are constants independent of n and k.*

If
$$\mathcal{G}_{m(n)}((t-x)^2, x) = \mathcal{O}((1-x^2)/n^2 + 1/n^4),$$

then there exists a constant C such that, for all $f \in C[-1, 1]$ and $x \in [-1, 1]$,

$$\mid f(x) - \mathcal{G}^+_{m(n)}(f, x) \mid \leq C\, \omega_2 \left(f, \sqrt{1-x^2}/n \right).$$

The last theorem can be applied when we used Jackson-Matsuoka or Fejér-
-Korovkin kernels. In particular, the nodes can be chosen as $x_{n-k} = \cos(k\pi/n)$,
$0 \leq k \leq n$.

Kacsó also constructed operators with equidistant nodes. Of course, more
than $n+1$ are needed.

Theorem 5.9.12 ([186]). *Let Δ_{n^2} be the partition given by the points $x_k = -1 + 2k/n^2$, $0 \leq k \leq n^2$. Let S_{n^2} be the operator constructed with these nodes. If*

$$\mathcal{G}_{m(n)}((t-x)^2, x) = \mathcal{O}((1-x^2)/n^2 + 1/n^4),$$

then there exists a constant C such that, for all $f \in C[-1, 1]$ and $x \in [-1, 1]$

$$\mid f(x) - (\mathcal{G}_{m(n)} \oplus S_{n^2})(f, x) \mid \leq C\, \omega_2 \left(f, \sqrt{1-x^2}/n \right).$$

Bibliography

[1] N.I. Akhieser and M.G. Krein, *On the best approximation of differentiable periodic functions by trigonometric sums*, Dokl. Akad. Nauk SSSR, 15 (1937), 107–112. FALTA

[2] V.V. Arestov, *On integral inequalities for trigonometric polynomials and their derivatives*, Math. USSR Izv. 18 (1982), 1–17.

[3] S.M. Asadov, *The comparison of two classes of functions in $L_{2,\alpha}[-1,1]$*, Akad. Nauk Azerbaidzhan, SSR Dokl. 28 (1972), 7–10.

[4] K.I. Babenko, *On some problems of the theory of approximations and numerical analysis*, Usp. Mat. Nauk, 40, No. 1, (1985), 3–27.

[5] V.F. Babenko and V.A. Kofanov, *Asymmetric approximations of classes of differentiable functions by algebraic polynomials*, Dokl. Akad. Nauk Ukr. SSR, Ser. A, 11 (1984), 3–5.

[6] V.F. Babenko and V.A. Kofanov, *Nonsymmetric approximation in the mean of differentiable functions by algebraic polynomials*, Anal. Math., 14 (3) (1988), 193–217,

[7] N.S. Baiguzov, *Some estimate for the derivatives of algebraic polynomials an application to numerical differentiation*, Mat. Zametki 5 (1969), 183–194.

[8] K. Balázs and T. Kilgore, *On the simultaneous approximation of derivatives by Lagrange and Hermite interpolation*, J. Approx. Theory, 60 (1990), 231–244.

[9] K. Balázs and T. Kilgore, *A discussion of simultaneous approximation of derivatives by Lagrange interpolation*, Numer. Funct. Anal. and Optim., 11 (1990), 225–237.

[10] K. Balázs and T. Kilgore, *On some constant in simultaneous approximation*, Internat. J. Math. Math. Sci. 18 (2) (1995), 279–286.

[11] K. Balázs, T. Kilgore and P. Vértesi, *Some remarks on Timan's approximation theorem. Approximation theory VI*, Proc. 6th Int. Symp., College Station/TX (USA) 1989, Vol. I (1989), 29–31.

[12] K. Balázs, T. Kilgore and P. Vértesi, *An interpolatory version of Timan's theorem on simultaneous approximation*, Acta Math. Hung., 57 (3-4) (1991), 285–290.

[13] M.S. Baouendi and C. Coulaoic, *Approximation polynomiale de fonctions C^∞ et analytiques*, Ann. Inst. Fourier (Grenoble), 21, No. 4, 149–173 (1971).

[14] N.K. Bari, *Trigonometric series*, Fizmatgiz, Moscow, 1961; English transl., Macmillan, New York; Pergamon Press, Oxford, 1964.

[15] N.K. Bari and S.B. Stechkin, *Best approximations and differential properties of two conjugate functions*, Trudy Mosk. Mat. Obshch., 5, 483–522 (1956).

[16] I.B. Bashmakova, *On the approximation of continuous functions by polynomial approximation methods on the segment*, Vestn. Leningr. Univ. 1978, No. 1 (Mat. Mekh. Astron. 1 (1978), 15–18.

[17] I.B. Bashmakova and V.N. Malozemov, *The approximation of continuous functions by algebraic polynomials* (Russian), Vestnik Leningrad. Univ. 28 (1968), 5–9.

[18] I.B. Bashmakova, *On an estimate for the approximation of continuous functions by algebraic polynomials*, Numerical methods in problems of mathematical physics, Inter Univ. Work Collect., Leningrad 1983, (1983), 5–11.

[19] H. Bavinck, *Jacobi Series and approximation*, Mathematical Center Tracts 39, Mathematisch Centrum, Amsterdam, 1972.

[20] H. Bavinck, *A special class of Jacobi series and some applications*, J. Math. Anal. Appl. 37 (1972), 767–797.

[21] H. Bavinck, *Approximation processes for Fourier-Jacobi expansions*, Applicable Anal. 5 (1976), 293–312.

[22] H. Berens, *Interpolationsmethoden zur Behandlung von Approximationsprozessen auf Banachräumen*, Lecture Notes in Math. 64, Springer Verlag, Berlin-Heidelberg-New York, 1968.

[23] H. Berens and Y. Xu, *On Bernstein-Durrmeyer polynomials with Jacobi weights*, In Approximation Theory and Functional Analysis, ed. by C.K. Chui, Academic Press, New York (1990), 25–46.

[24] H. Berens, Y. Xu, *On Bernstein-Durrmeyer polynomials with Jacobi weights: the case $p = 1$ and $p = \infty$*, In Approximation, Interpolation and Summation, eds. S. Barron et al., Jerusalem, Weizmann Science Press, 1991, 51–62.

[25] S. Bernstein, *Sur l'approximation des fonctions continues par des polynômes*, Comptes Rendus, 152 (1911), 502–504.

[26] S.N. Bernstein, *Sur l'ordre de la meilleure approximation des fonctions continues par des polynômes de degré donné*, Mémoires Couronnés Acad. Roy. Belg. (2) 4 (1912), 1–104.

[27] S.N. Bernstein, *On the best approximation of continuous functions by polynomials of a given degree*, Works, Vol. I, 11–104.

[28] S.N. Bernstein, *Sur une modification de la formula d'interpolation de La-grange*, Comm. Sot. Math. Kharkov 5 (1931), 49–57 (Works, T. II, 130–140).

[29] S.N. Bernstein, *On the limit relations between the constant of the theory of best approximation*, Dokl. Akad. Nauk, 57 (1947), 3–5.

[30] S.N. Bernstein, *On the approximation of continuous functions by polynomi-als*, in: Collected Works, Academy of Sciences of the USSR, Vol. 1 (1952), 8–10.

[31] P. Binev, $\mathcal{O}(n)$ *bounds of Whitney constants*, C. R. Acad. Bulgare Sci. 38 (1985), 1315–1317.

[32] R. Bojanic, *A note on the degree of approximation to continuous functions*, L'Enseignement Math. 15 (1969), 43–51.

[33] R. Bojanic, R. DeVore, *On polynomials of best one-sided approximation*, L'Enseignement Mathématique, Tom XII, fasc. 3, 1966, 139–164

[34] R. Bojanic, R. DeVore, *A proof of Jackson's theorem*, Bull. Amer. Math. Soc. 75 (1969), 364–367.

[35] A. Boos, J. Cao and H.H. Gonska, *Approximation by Boolean sums of posi-tive linear operators. V: On the constants in Devore-Gopengauz-type inequal-ities*, Calcolo, 30 (3) (1993) 289–334.

[36] E. Borel, *Leçons sur les Fonctions de Variables Réelles et les Développe-ments en Séries de Polynômes*, Paris, 1905, pp. 82–92.

[37] Yu.A. Brudnyi, *Approximations by Entire Function in Regions Exterior to a Segment and to a Semiaxis*, Dokl. Akad. Nauk SSSR, 124, No. 4 (1959).

[38] Yu.A. Brudnyi, *A generalization of a theorem of A.F. Timan*, Dokl. Akad. Nauk SSSR, 188 (6) (1963) 1237–1240.

[39] Yu.A. Brudnyi, *On a theorem of local best approximations*, Uchen. Zap. Kazan. Gos. Univ., 124 (1964), no. 6, 43–49.

[40] Yu.A. Brudnyi, *Approximation of functions by algebraic polynomials*, Izv. Akad. Nauk SSSR, Ser. Mat. 32, 1968, 780–787.

[41] Yu.A. Brudnyi and I.E. Gopengauz, *The measure of the point set of maximal deviation*, Izv. Akad. Nauk SSSR, Ser. Mat., 24, No. 1, 129–144 (1960).

[42] Yu.A. Brudnyi and I.E. Gopengauz, *Approximation by piecewise-polynomial functions*, Izv. Akad. Nauk SSSR, Ser. Mat., 27, No. 4, (1963), 723–746.

[43] H. Burkill, *Cesaro-Perron almost periodic functions*, Proc. London Math. Soc. 3 (1952), 150–174.

[44] P.L. Butzer, *Legendre transform methods in the solution of basic problems in algebraic approximation*, in Functions, Series, Operators, Vol. 1, (B. Sz.-Nagy and J. Szabados, eds.), North-Holland, Amsterdam, 1983, 277–301.

[45] P.L. Burzer and H. Berens, *Semi-Groups of Operators and approximation*, Grundlehren Math. Wiss., 145, Springer, Berlin, 1967.

[46] P.L. Butzert and S. Pawelke, *Ableitungen von trigonometrischen Approximationsprozessen*, Acta Sci. Math. Szeged 28 (1967), 173–183.

[47] P.L. Butzer, E. Görlich and K. Scherer, *Introduction to Interpolation and approximation*, (unpublished), 1973.

[48] P.L. Butzer, S. Jansche and R.L. Stens, *Functional analytic methods in the solution of the fundamental theorems of best algebraic approximation*, in Approximation Theory (G.A. Anastassiou, ed.) Lectures Notes in Pure and Appl. Math. 138, Dekker, New York, 1992, 151–205.

[49] P.L. Butzer and B. Sz. Nagy (eds.), *Linear Operators and Approximation II*, Proceeding of the Conference held in Oberwolfach, ISNM Vol. 25, Birkhäuser (1974).

[50] P.L. Butzer and K. Scherer, *Approximationsprozesse und Interpolationsmethoden*, B. I. Hochschulskripten 826/826a, Bibliographisches Institut, Mannheim 1968.

[51] P.L. Butzer and K. Scherer, *On the fundamental approximation theorem of D. Jackson. S.N. Bernstein and theorem of M. Zamaski and S.B. Stechkin*, Aequationes Math., 3 (1969), 170–185.

[52] P.L. Butzer and K. Scherer, *Über die Fundamentalsätze der klassichen Approximationstheorie in abstrakten Räumen*, in Abstract Spaces and approximation, P.L. Butzer et al., Internat. Ser. Numer. Math. 10, Birkhäuser, Basel, 1969, 113–125.

[53] P.L. Butzer and K. Scherer, *Approximation theorems for sequences of commutative operators in Banach spaces*, in Constructive Function Theory, Publ. House Bulgarian Acad. Sci., 1972, 137–146.

[54] P.L. Butzer and K. Scherer, *Jackson and Bernstein-type inequalities for families of commutative operators in Banach spaces*, J. Approx. Theory, 5 (1972), 308–342.

[55] P.L. Butzer and R.L. Stens, *Chebyshev transform methods in the theory of best algebraic approximation*, Abh. Math. Sem. Univ. Hamburg 45 (1976), 165–190.

[56] P.L. Butzer and R. L. Stens, *The operational properties of the Chebyshev transform I: General properties*, Funct. Approx. Comment Math., 5 (1977) 129–160.

[57] P.L. Butzer and R.L. Stens, *The operational properties of the Chebyshev transform II: Fractional derivatives*, Theory of Approximation of Functions, S.B. Stechkin et al., eds., Nauka, Moscow, 1977, 49–61.

[58] P.L. Butzer, R.L. Stens, and M. Wehrens, *Approximation by algebraic convolution integrals*, in Approximation Theory and Functional Analysis (J. Proll, ed.), North-Holland, Amsterdam, 1979, 71–120.

[59] P.L. Butzer, R.L. Stens and M. Wehrens, *Saturation classes of the Cesàro and Abel-Poisson means of Fourier-Legendre series*, Acta Math. Hungar. 33 (1979), 19–35.

[60] P.L. Butzer, R.L. Stens and M. Wehrens, *Higher order moduli of continuity based on the Jacobi translation operator and best approximation*, C. R. Math. Rep. Acad. Sci. Canada 2 (1980), 83–88.

[61] P.L. Butzer and M. Wehrens, *Bernstein-type polynomials having $O(n^{-2})$ as saturation order*, Analysis, 1 (1981), 163–169.

[62] J.-D. Cao and H.H. Gonska, *Approximation by Boolean sums of positive linear operators*, Rend. Mat. Appl., VII. Ser. 6, No.4, (1986), 525–546.

[63] J.-D. Cao and H.H. Gonska, *Approximation by Boolean Sums of Positive Linear Operators. II. Gopengauz-Type Estimates*, J. Approx. Theory 57 (1989), 77–89.

[64] J.-D. Cao and H.H. Gonska, *Approximation by Boolean sums of positive linear operators III: estimates for some numerical approximation schemes*, Numer. Funct. Anal. Optim. 10 (1989), 643–672.

[65] J.-D. Cao and H.H. Gonska, *Pointwise estimate for modified positive linear operators*, Portugaliae Math., 46 (4) (1989), 401–430.

[66] J.-D. Cao and H.H. Gonska, *Computation of DeVore-Gopengauz-type approximants*, in Approximation Theory VI (C.K. Chui et al.), Academic Press, New York, 1989, 117–120.

[67] J.-D. Cao and H.H. Gonska, *Approximation by Boolean sums of linear operators. Telyakovskii-type estimates*, Bull. Austral. Math. Soc., 42 (1990), 253–266.

[68] J.-D. Cao and H.H. Gonska, *On Butzer's problem concerning approximation by algebraic polynomials*, in G. Anastassiou, ed., Approximation Theory (Proc. Sixth Southeastern Approximation Theorists Annual Conference, Memphis / TN 1991), Marcel Dekker (New York, 1992), 289–313.

[69] J.-D. Cao and H.H. Gonska, *Pointwise estimates for higher order convexity preserving polynomial approximation*, J. Austral. Math. Soc. Ser. B, 36 (1994), 213–233.

[70] J.-D. Cao and H.H. Gonska, *Solution of Butzer's problem (linear form) and some algebraic polynomials operators with saturation order $\mathcal{O}(n^{-2})$*, in Proc. of the Third International Colloquium on Numerical Analysis, Plovdiv, 1994 (D. Bainov and V. Covachev, eds.), VSP, Zeist, The Netherlands, 1995, 37–41.

[71] J.-D. Cao and H.H. Gonska, *New Teliakovskii-type estimates via the Boolean sum approach*, Bull. Austral. Math. Soc., 54 (1996), 131–146.

[72] J.-D. Cao, H.H. Gonska and D.P. Kacsó, *Simultaneous approximation by discretized convolution-type operators*, D.D. Stancu (ed.) et al., Approximation and optimization. Proceedings of ICAOR: international conference,

Cluj-Napoca, Romania, July 29–August 1, 1996. Vol. I. Cluj-Napoca: Transilvania Press, (1997), 203–218.

[73] J.-D. Cao, H.H. Gonska and H.J. Wenz, *Approximation by boolean sums of positive linear operators VII: Fejér-Korovkin kernels of higher order*, Acta Math. Hungar. 73 (1-2) (1996), 71–85.

[74] B.P.S. Chauhan, *A proof of Telyakoski-Gopengaus's theorem through an interpolation process of S.N. Bernstein*, Indian J. Pure Appl. Math., 13 (12) (1982), 1433–1438.

[75] B.P.S. Chauhan, *On the theory of approximation of continuous functions of interpolatory polynomials*, Indian J. Pure Appl. Math., 14 (8) (1983), 1041–1048.

[76] M.M. Chawla, *Approximation by non-negative algebraic polynomials, Nordisk Tidskr. Informationsbobandling*, BIT Numerical Mathematics 10 (1970), 243–248.

[77] S. Csibi, *Notes on de la Vallée-Poussin's approximation theorem*, Acta Math. Acad. Sci. Hung. 7 (1956), 435–439.

[78] J. Czipszer and G. Freud, *Sur l'approximation d'une fonction périodique et de ses dérivées successive par un polynôme trigonométrique et par ses dérivées successive*, Acta Math., 99 (1958), 33–51.

[79] R. Dahlhaus, *Pointwise approximation by algebraic polynomials*, J. Approx. Theory, 57 (1989), 274–277.

[80] F. Dai, Z. Ditzian, and S. Tikhonov, *Sharp Jackson inequalities,* J. Approx. Theory, 151 (2008), 86–112.

[81] C. de la Vallée-Poussin, *Sur la convergence des formules d'interpolation entre ordonnées equidistant*, Bull. Acad. Belgique, 1908

[82] C. de la Vallée-Poussin, *Sur l'approximation des fonctions d'une variable réelle et de leurs dérivées par des polynômes et des suites limitées de Fourier*, Bull. de l'Acad. Royale Belgique, Classe des Sciences (1908), 193-254.

[83] C. de la Vallée-Poussin, *Sur les polynômes d'approximation et la représentation approchée d'un angle*, Bull. Acad. Belgique, 1910.

[84] C. de la Vallée-Poussin, *Sur le maximum du module de la dérivée d'une expression trigonométrique d'ordre et de module bornés*, Comptes Rendus, 166 (1918), 843–846.

[85] C. de la Vallée-Poussin, *Leçons sur l'approximation des fonctions d'une variable réelle*, Paris, 1919.

[86] R. DeVore, *On Jackson's theorem*, J. Approx. Theory, 1968, 314–318.

[87] R. DeVore, *Pointwise approximation by polynomials and splines*, Proc. Conf. on Approximation of Functions, Kaluga, S.B. Stechkin et al. eds., Nauka, Moscow, 1977, 132–141.

[88] R. DeVore, *Degree of approximation*, Approximation Theory II, edited by G.G. Lorentz et al., Academic Press, New York, 1976, 117–161.

[89] R. DeVore, $L_p[-1, 1]$ *approximation by algebraic polynomials, Linear spaces and Approximation*, (ed. P.L. Butzer and B. Sz. Nagy), Birkhäuser Verlag, Basel, 1978, 397–406.

[90] R.A. DeVore, D. Leviatan, X. Ming Yu, *Polynomial Approximation in L_p, $(0 < p < 1)$*, Constr. Approx., 8 (1992), 187–201.

[91] R.A. DeVore and X.M. Yu., *Pointwise estimates for monotone polynomial approximation*, Constr. Approx., 4 (1985), 323–331.

[92] Z. Ditzian, *On global inverse theorems for combinations of Bernstein polynomials*, J. Approx. Theory, 26 (1979), 277–292.

[93] Z. Ditzian, *On interpolation on $L_p[a, b]$ and weighted Sobolev classes*, Pacific J. Math., 90 (1980), 307–323.

[94] Z. Ditzian, *Interpolation theorems and the rate of convergence of Bernstein polynomials*, Approximation Theory III, ed. E.W. Cheney, Academic Press, 1980, 341–347.

[95] Z. Ditzian, *Polynomials of best approximation in $C[-1, 1]$*, Israel J. Math., 52 (4) (1985), 341–354.

[96] Z. Ditzian, *On the Marchaud-type inequality*, Proc. Amer. Math. Soc. 103 (1988), 198–202.

[97] Z. Ditzian, *A note on simultaneous polynomial approximation in $L_p[-1, 1]$, $0 < p < 1$*, J. Approx. Theory, 82 (1995), 317–319.

[98] Z. Ditzian, V.H. Hristov and K.G. Ivanov, *Moduli of smoothness and K-functionals in L_p, $0 < p < 1$*, Constr. Approx. 11, 67–83.

[99] Z. Ditzian and D. Jiang, *Approximation of functions by polynomials in $C[-1, 1]$*, Canadian J. Math., 44 (5) (1992), 924–940.

[100] Z. Ditzian, D. Jiang and D. Leviatan, *Simultaneous polynomial approximation*, SIAM J. Math. Anal., 24(6) (1993), 1652–1661.

[101] Z. Ditzian, D. Jiang, D. Leviatan, *Inverse theorem for the best approximation in L_p, $0 < p < 1$*, Proc. Amer. Math. Soc., 120 (1994), 151–153.

[102] Z. Ditzian, V. Totik, *Moduli of smoothness*, Springer, New York, 1987.

[103] E.P. Dolzhenko and E.A. Sevast'yanov, *On the dependence of the properties of functions on the rate of their approximation by polynomials*, Izv. Akad. Nauk SSSR, Ser. Mat., 42 (2) (1978), 270–304.

[104] E. Dynkin, *A constructive characterization of Sobolev and Besov spaces*, Trudy Math. Inst. Steklov, 155 (1981), 39–74.

[105] A.S. Dzafarov, *Certain direct and inverse theorems in the theory of best approximations of functions by algebraic polynomials*, Dokl. Akad. Nauk SSSR 187 (1969), 719–722.

[106] G.A. Dzyubenko, *Copositive pointwise approximation*, Ukranian Mat. Zh., 48 (1996), 326–334.

[107] V.K. Dzyadyk, *The constructive characteristic of functions satisfying* Lip$_\alpha$ *condition* $(0 < \alpha \leq 1)$ *on a finite segment of the real axis*, Izv. Akad. Nauk SSSR Ser. Mat. 20 (1956), 623–642.

[108] V.K. Dzyadyk, *Continuation of functions satisfying the Lipschitz condition in the L metric*, Mat. Sb. 40 (82) (1956), 239–242.

[109] V.K. Dzjadyk, *Approximation of functions by ordinary polynomials on a finite interval of the real axis*, Izv. Akad. Nauk SSSR Ser. Mat. 22 (1958), 337–354.

[110] V.K. Dzyadyk, *A further strengthening of Jackson's theorem on the approximation of a continuous functions by ordinary polynomials*, Dokl. Akad. Nauk 121 (1958), 403–406.

[111] V.K. Dzyadyk, *Investigations in the theory of approximation of analytic functions carried out at the Mathematics Institute*, Academy of Sciences of the Ukranian SSR, Ukrainskii Math. Zhurnal, 21 (2) (169), 173–192.

[112] V.K. Dzyadyk, *Introduction to the Theory of Uniform Approximation of Functions by Polynomials*, Nauka, Moscow, 1977.

[113] V.K Dzyadyk and I.A. Shevchuk, *Theory of Uniform Approximation of Functions by Polynomials*, Walter de Gruyter, Berlin, Germany, 2008.

[114] P. Erdös, *On some convergence properties of the interpolation polynomials*, Annals of Math., 44 (1943), 330–337.

[115] J. Favard, *Sur les meilleurs procédés d'approximation de certaines classes de fonctions par des polynômes trigonométriques*, Bull. Sci. Math. 61 (1937), 209–224, 243–256.

[116] J. Favard, *Sur l'approximation des fonctions*, Bull, de Sc. Math., LXII (1938), 338–351.

[117] R.P. Feinerman and D.J. Newman, Polynomial Approximation, Williams and Wilkins, Baltimore, 1974.

[118] L. Fejér, *Die Abschätzung eines Polynoms in einem Intervalle*, Math. Z., 32 (1930), 426-457.

[119] M. Felten, *A modulus of smoothness based on an algebraic addition*, Aequationes Math., 54(1-2) (1997), 56–73.

[120] M. Felten, *Characterization of best algebraic approximation by an algebraic modulus of smoothness*, J. of Approx. Theory, 89 (1) (1997), 1–25.

[121] S.D. Fisher, *Best approximation by polynomials*, J. Approx. Theory 21 (1977), 43–59.

[122] G. Freud, *Über gleichzeitige Approximation einer Funktion und ihrer Derivierten*, Math. Nachrichten, Wien, 47/49 (1957), 36–37.

[123] G. Freud, *Über die Approximation reeller stetigen Functionen durch gewöhnliche Polynome*, Math. Ann. 137 (1959), 17–25.

[124] G. Freud, *Über ein Jacksonsches Interpolationsverfahren*, in: On Approximation Theory, Proc. Conference in Oberwolfach (eds. P.L. Butzer and J. Korevaar), ISNM Vol. 5 (Birkhäuser, 1963), 227–232.

[125] G. Freud, *On approximation by interpolatory polynomials*, Math. Lapok, (18) (1967), 61–64.

[126] G. Freud and A. Sharma, *Some good sequences of interpolatory polynomials*, Canad. J. Math. 26 (1974), 233–246.

[127] G. Freud and A. Sharma, *Addendum: Some good sequences of interpolation polynomials*, Canad. J. Math. 29 (1977), 1163–1166.

[128] G. Freud and P. Vértesi, *A new proof of Timan's approximation theorem*, Studia Sci. Math. Hungar, 2(1967), 403–414.

[129] A.L. Fuksman, *Structural characteristic of functions such that $E_n(f-1,1) \leq Mn^{-(k+\alpha)}$*, Uspekhi Math. Nauk, 20 (1965), 187–190.

[130] C.C. Ganser, *Modulus of Continuity conditions for Jacobi series*, J. Math. Anal. Appl. 27 (1969), 575–600.

[131] M.I. Ganzburg, *Polynomial interpolation, an L-function, and pointwise approximation of continuous functions*, J. Approx. Theory, 153 (1) (2008), 1–18.

[132] I.M. Ganzburg, A.F. Timan, *Linear processes of approximation by algebraic polynomials to functions satisfying a Lipschitz condition*, Izv. Akad. Nauk SSSR Ser. Mat., 22 (6) (1958), 771–810.

[133] A.L. Garkavi, *Simultaneous approximation of a periodic function and its derivatives by trigonometric polynomials*, Izv. Akad. Nauk SSSR Ser. Mat. 24 (1960), 103–128.

[134] G. Gasper, *Banach algebras for Jacobi series and positivity of a kernel*, Ann. Math. (2) 95 (1972), 261–280.

[135] I. Gavrea, *Approximation by positive linear operators*, Anal. Numér. Théor. Approx., 24 (1995), 109–115.

[136] I. Gavrea, *The approximation of the continuous functions by means of some linear positive operators*, Resultate Math., 30 (1996), 55–66.

[137] I. Gavrea, H.H. Gonska and D. Kacsó, *Positive linear operators with equidistant nodes*, Computers Math. Applic., 32 (8), 1996, 23–32.

[138] I. Gavrea, H.H. Gonska and D.P.D. Kacsó, *Variation on Butzer's problem: Characterization of the solutions*, Computers Math. Applic., 34 (99) (1997), 51–64.

[139] I. Gavrea, H.H. Gonska, and D.P. Kacsó, *On linear operators with equidistant nodes: negative results*, Rend. Circ. Mat. Palermo (2) Suppl. 52 (1998), 445–454.

[140] I. Gavrea, H.H. Gonska, and D.P. Kacsó, *On discretely defined positive linear polynomial operators giving optimal degrees of approximation*, Rend. Circ. Mat. Palermo (2) Suppl. 52 (1998), 455–473.

[141] A.O. Gelfond, *On uniform approximation by polynomials with integer coefficients*, Uspekhi Mat. Nauk, 1 (63), 1955, 41–65.

[142] J. Gilewicz, Yu.V. Kryakin, and I.A. Shevchuk, *Boundedness by 3 of the Whitney interpolation constants*, J. Approx. Theory 119 (2002), 271–290.

[143] M. v. Golischek, *Die Ableitungen der algebraischen Polynome bester Approximation*, in Approximation Theory, Z. Ciesielski et al., ed., Reidel, Dordrecht, and PWN – Polish Scientific Publishers, Warszawa, 1975, 71–86.

[144] H.H. Gonska, *On approximation in spaces of continuous functions*, Bull. Austral. Math. Soc. 28 (1983), 411–432.

[145] H.H. Gonska, *Modified Pichugov-Lehnhoff operators*, in Approximation Theory V, Proceedings International Symposium, College Station, Texas, 1986 (C.K. Chui, ed.), Academic Press, New York, 1986, 355–358.

[146] H.H. Gonska and J.-d. Cao, *On Butzer's problem concerning approximation by algebraic polynomials*, in Approximation Theory, Proceedings Sixth Southeastern Approximation Theorists Annual Conference, Memphis, TN, March 1991 (G. Anastassiou, ed.), Marcel Dekker, New York, 1992, 289–313.

[147] H. Gonska and E. Hinnemann, *Punktweise Abschätzungen zur Approximation durch algebraische Polynome*, Acta Math. Hungar., 46 (3-4) (1985), 243–254.

[148] H.H. Gonska, D. Leviatan, I.A. Shevchuk and H.-J. Wenz, *Interpolatory pointwise estimates for polynomial approximation*, Constr. Approx., 16 (2000), 603–629.

[149] H.H. Gonska, X.-L. Zhou, *Polynomial approximation with side condition: recent results and open problems*, Proc. 1st Intern. Colloq. Numerical Anal., (D. Bainov and V. Covachev, eds.), VSD Inter. Sci., Zeis/The Netherlands (1993), 61–71.

[150] I.E. Gopengauz, *A.F. Timan's theorem regarding the approximation of functions by polynomials on a finite segment*, Mat. Zametki, 1 (2) (1967), 163–172.

[151] I.E. Gopengauz, *A question concerning the approximation of functions on a segment and a region with corners*, Teor. Funktsii FunktsionaL Anal i Prilozhen, 4 (1967), 204–210.

[152] I.E. Gopengauz, *Pointwise estimate of the error of interpolation with multiple nodes in the ends of a closed interval*, Mat. Zametki, 51 (1992), 55–61.

[153] G. Grünwald, *Über Divergenzerscheinungen der Lagrangeschen Interpolationspolynome stetiger Funktionen*, Annals of Mathematics, (2) (37) (1936), 908–918.

[154] G. Grünwald, *On a convergence theorem for the Lagrange interpolation polynomials*, Bull. Amer. Math. Soc. 47 (1941), 271–275.

[155] E.G. Guseinov and N.A. ll'yasov, *Differential and smoothness properties of continuous functions*, Math. Notes, 1977, 931–936.

[156] M. Hasson, *Derivatives of the algebraic polynomials of best approximation*, J. Approx. Theory 29 (1980), 91–102.

[157] M. Hasson, *The sharpness of Timan's theorem on differentiable functions*, J. Approx. Theory, 3, 264–274 (1982)

[158] M. Heilmann, *On simultaneous approximation by optimal algebraic polynomials*, Results in Mathematics, 16 (1989), 77–81.

[159] E. Hinnemann and H.H. Gonska, *Generalization of a theorem of DeVore*, Approximation Theory IV, C.K. Chui, L.L. Schumaker and J.W. Ward, Academic Press, New York, 1983, 527–532.

[160] P.G. Hoel, *Certain problems in the theory of closest approximation*, Amer. J. Math, 57 (1935), 891–901.

[161] V.I. Ivanov, *Direct and converse theorems of the theory of approximation in the metric of L_p for $0 < p < 1$*, Math. Notes, 18 (1975), 872–982.

[162] V.I. Ivanov, *Certain inequalities in various metrics for trigonometric polynomials and their derivatives*, Mathematical Notes, (4) 18 (1975), 880–885.

[163] K.G. Ivanov, *Direct and converse theorems for the best algebraic approximation in $C[-1,1]$ and $L_p[-1,1]$*, Comptes Rendus Aca. Bulg. Sic, 33 (10) (1980), 1309–1312.

[164] K.G. Ivanov, *Direct and converse theorems for the best algebraic approximation in $C[-1,1]$ and $L_p[-1,1]$*, Coll. Mat. Soc. János Blyai (1980), 675–682.

[165] K.G. Ivanov, *Some characterizations of the best algebraic approximation in $L_p[-1,1]$ $(1 \le p \le \infty)$*, C. R. Bulgare Sci. 34 (1981), 1229–1232.

[166] K.G. Ivanov, *On a new characteristic of functions*, Serdica 8 (1982), 262–279.

[167] K.G. Ivanov, *On a new characteristic of functions. II. Direct and converse theorems for the best algebraic approximation in $C[-1,1]$ and $L_p[-1,1]$*, PLISKA, Stud. Math. Bulgar. 5 (1983), 151–163.

[168] K.G. Ivanov, *On the rates of convergence of two moduli functions*, PLISKA, Stud. Math. Bulgar. 5 (1983), 97–104.

[169] K.G. Ivanov, *A constructive characteristic of the algebraic approximation in $L_p[-1,1]$*, Constructive Functions Theory '81, Sofia, 1983, 357–367.

[170] K.G. Ivanov, *On the dependence of the differential properties of a function of its best algebraic approximation*, Serdica, 10 (1984), 184–191.

[171] K.G. Ivanov, *On the behaviour of two moduli functions*, C. R. Bulgare Sci. 38 (1985), 539–542.

[172] K.G. Ivanov, *On the behaviour of two moduli functions. II*, Serdica, 12 (1986), 196–203.

[173] K.G. Ivanov, *A characterization of weighted Petree K-functionals*, J. Approx. Theory, 56 (1989), 185–211.

[174] K. Ivanov and V. Takev, $\mathcal{O}(n \ln n)$ *bounds of constants of H. Whitney*, C. R. Acad. Bulgare Sci. 38 (1985), 1129–1131.

[175] D. Jackson, *Über die Genauigkeit der Annäherung stetiger Funktionen durch ganze rationale Funktionen gegebenen Grades und trigonometrische Summen gegebener Ordnung*, Dissertation, Göttingen, 1911.

[176] D. Jackson, *On approximation by trigonometric sums and polynomials*, Trans. Amer. Math. Soc, 13 (1912), 491–515.

[177] D. Jackson, *The general theory of approximation by polynomials and trigonometric sums*, Bull. Amer. Math. Soc., (1921), 415–431.

[178] D. Jackson, *On the convergence of certain trigonometric and polynomial approximations*, Trans Amer. Math. Soc., 22 (1921), 158–166.

[179] D. Jackson, *The Theory of Approximation, Vol. XI. Amer. Math. Soc. Colloquium Publications, New York, 1930.*

[180] He Jiaxing, *Convergence order of an interpolation process by Bernstein polynomial*, Acta Math. Hungar., 53 (1989), 281–287.

[181] He Jiaxing, *On an interpolation process of S.N. Bernstein*, J. Approx. Theory, 60 (1990), 123–132.

[182] He Jiaxing and Ye Jichang, *On an interpolation polynomial of S.N. Bernstein type*, Acta Mathematica Hungarica, 70 (4), 1996, 293–303.

[183] He Jiaxing, Zhang Yulei and Li Songtao, *On a new interpolation process of S.N. Bernstein*, Acta Mathematica Hungarica, 73 (4), 1996, 327–334.

[184] He, Jiaxing and Li, Xiaoniu, *A new third S.N. Bernstein interpolation polynomial*, Chin. Q. J. Math. 13 (4) (1998), 10–16.

[185] J. Junggebutrh, K. Scherer and W. Trebels, *Zur besten Approximation auf Banachräumen mit Anwendungen auf ganze Funktionen*, Forschungsberichte des Landes Nordrhein-Westfalen, Nr. 2311 (1973), Westdeutscher Verlag, Opladen, 51–75.

[186] D. Kacsó, *Discrete Jackson-type operators via a Boolean sum approach*, J. Comp. Anal. Appl., 3 (4) 2001, 399–413.

[187] B.A. Khalilova, *Fourier-Jacobi's coefficient and approximation of functions to ultrasperical polynomials* (Russian), Izv. Akad. Nauk Azerbatdzhan. SSR Ser. Fiz.-Tehn. Mat. Nauk 2 (1973), 87–94.

[188] B.A. Khalilova, *On some estimations for polynomials*, Izv. Akd. Nauk Azerbaizhan SSR Ser. Fiz. Mat. Nauk 2 (1974), 46–55.

[189] L.B. Khodak, *Structural characteristics of some classes of functions in the metric of L_p for $0 < p < 1$*, Ukrainian Mathematical Journal, (3) 32 (1980), 287–293.

[190] L.B. Khodak, *Approximation of functions by algebraic polynomials in the metric of L_p for $0 < p < 1$*, Mathematical Notes, 30 (3) (1981), 649–655.

[191] T. Kilgore, *An elementary simultaneous approximation theorem*, Proc. Amer. Math. Soc., 118 (2), 1993, 529–536.

[192] T. Kilgore, *On the simultaneous approximation of functions and their derivatives*, G.A. Anastassiou (ed.), Applied mathematics reviews. Volume 1. Singapore: World Scientific. (2000), 69–118

[193] T. Kilgore and J. Prestin, *On the order of convergence of simultaneous approximation by Lagrange-Hermite interpolation*, Math. Nachr., 150 (1991), 143–150.

[194] T. Kilgore and J. Prestin, *A theorem of Gopengauz type with added interpolatory conditions*, Numer. Funct. Anal. Optimization, 15 (7-8) (1994), 859–868.

[195] T. Kilgore and J. Prestin, *Pointwise Gopengauz estimate for interpolation*, Annales Univ. sci. Budapestinensis, Sectio Computatorica, 16 (1996), 253–261.

[196] T. Kilgore and J. Szabados, *On approximation of a function and its derivatives by a polynomial and its derivatives*, Approximation Theory Appl., 10 (3) (1994), 93–103.

[197] O. Kis, *On some interpolation process of S.N. Bernstein*, Acta Math. Acad. Sci. Hungar., 24 (3-4) (1973), 353–361.

[198] O. Kis and J. Szabados, *On some de la Vallée-Poussin type discrete linear operators*, Acta Math. Hungar. 47 (1986), 239–260.

[199] O. Kis, P. Vértesi, *On a new interpolation process*, Annales Univ. Sci. Budapest, Sectio Math., 10 (1967), 117–128.

[200] V.A. Kofanov, *Best uniform approximation of differentiable functions by algebraic polynomials*, Math. Notes, 27 (1980), 190–195.

[201] V.A. Kofanov, *Approximation in the mean of classes of differentiable functions by algebraic polynomials*, Sov. Math., Dokl., 25 (1982), 224–226.

[202] V.A. Kofanov, *Approximation of the classes of differentiable functions by algebraic polynomials in mean*, Izv. Akad. Nauk SSSR, Ser. Mat., 47 (5) (1983), 1078–1090.

[203] V.A. Kofanov, *Approximation by algebraic polynomials of classes of functions which are fractional integrals of integrable functions*, Anal. Math., 13 (3) (1987), 211–229.

[204] S.V. Koniagin, *Bounds on the derivatives of polynomials*, Dokld. Akad. Nauk SSSR, 243 (1978), 1116–118.

[205] K.A. Kopotun, *Unconstrained and convex polynomial approximation in $C[-1, 1]$*, Approx. Theory Appl., 11(2) (1995), 41–58.

[206] K. Kopotun, *Simultaneous approximation by algebraic polynomials*, Constructive Approximation, 12 (1), (1996), 67–94.

[207] N.P. Korneichuk, *The exact constant in D. Jackson's theorem on best uniform approximation of continuous periodic functions*, Dokl. Akad. Nauk SSSR, 145 (1962), 514–515.

[208] N.P. Korneichuk, *Best approximation of continuous functions*, Izv. Akad. Nauk SSSR, Ser. Matem., 277, 29–44 (1963).

[209] N.P. Korneichuk, *The best approximation on an interval of classes of functions with bounded rth derivative by finite dimensional subspaces*, Ukrain. Mat. Zh. 31 (1979), 23–31. Ukrainian Math. J. 31 (1979), 16–23.

[210] N.P. Korneichuk, *Exact Constants in Approximation Theory*, Encyclopedia of Mathematics and its Applications, Cambridge University Press, 1991.

[211] N.M. Korneichuk and A.I. Polovina, *Approximation of continuous differentiable functions by algebraic polynomials on a segment*, Dokl. Akad. Nauk SSSR 166 (1966), 281–283.

[212] N.P. Korneichuk and A.I. Polovina, *Algebraic-polynomial approximation of functions satisfying a Lipschitz condition*, Mat. Zametki, 9 (4) (1971), 441–447.

[213] N.P. Korneichuk and A.I. Polovina, *On approximation of continuous functions by algebraic polynomials*, Ukr. Mat. Zh., (1972), 269–278.

[214] V.G. Krotov, *On differentiability of functions in L^p*, $0 < p < 1$, Math. USSR Sbornik, 45 (1983), 101–119.

[215] Yu.V. Kryakin, *On the Whitney constants*, Mat. Zametki 46 (1989), 155–157.

[216] Yu.V. Kryakin, *On the constant of H. Whitney in L_p*, $1 \leq p \leq \infty$, Math. Balkanica (N.S.) 4 (1990), 158–174.

[217] Yu.V. Kryakin, *On the exact constants in Whitney's theorem*, Mat. Zametki 54 (1993), 34–51

[218] Yu.V. Kryakin, *On Whitney's theorem and constant*, Russian Acad. Sci. Sb. Math., 2 (81) (1995), 281–295.

[219] Yu.V. Kryakin, *On functions with a bounded nth difference*, Izv. Ross. Akad. Nauk, Ser. Mat. 61 (1997), 95–100.

[220] Yu.V. Kryakin and L.G. Kovalenko, *On the Whitney constants for classes L_p*, $1 \leq p < \infty$, Izv. Vysh. Uchebn. Zaved. Mat. 1 (1992), 69–77.

[221] Yu.V. Kryakin and M.D. Takev, *Interpolation Whitney constants*, Ukrainian Mathematical Journal, 47 (8) (1995), 1188–1194.

[222] N.M. Kryloff, *Sur quelques formules d'interpolation généralisée*, Bull. Sci. Math., 2 (41) (1917), 309–320.

[223] N.L. Labunetz, *On the question of approximating functions that are continuous on a segment by truncated de la Vallée-Poussin sums*, Ivuz, 8 (255) (1983), 32–33.

[224] E. Landau, *Über die Approximation einer stetigen Funktion durch eine ganze rationale Funktion*, Rendiconti Cir. Mat. Palermo, 25 (1908), 337–345.

[225] G.K. Lebed, *Inequalities for polynomials and their derivatives*, Dokl. Akad. Nauk SSSR 117 (1958), 570–572.

[226] G.K. Lebed, *Some questions of approximation of functions of one variable by algebraic polynomials*, Dokl. Akad. Nauk SSSR, 118 (2) (1958), 239–242.

[227] G.K. Lebed, *Some inequalities for trigonometric and algebraic polynomials and their derivatives*, Tr. Mat. Inst. Steklova 134, (1975), 142–160.

[228] H. Lebesgue, *Sur l'approximation des fonctions*, Bull. Sci. Math., (2) (22) (1898), 278–287.

[229] H. Lebesgue, *Sur la représentation approchée des fonctions*, Rendiconti Cir. Mat. Palermo, 26 (1908), 325–328.

[230] H.G. Lehnhoff, *A simple proof of A F. Timan's theorem*, J. Approx. Theory, 1983, 172–176.

[231] H.G. Lehnhoff, *A new proof of Teliakovskii's theorem*, J. Approx. Theory, 38 (1983), 177–181.

[232] M. Lerch, *About the main theorem of the theory of generating functions* (in Czech), Rozpravy Ceske Akad. 33 (1892), 681–685.

[233] M. Lerch, *Sur un point de la théorie des fonctions génératrices d'Abel*, Acta Math. 27 (1903), 339–351

[234] D. Leviatan, *The behavior of the derivatives of the algebraic polynomials of best approximation*, J. Approx. Theory, 35 (1982), 169–176.

[235] W. Li, *On Timan type theorems in algebraic polynomial approximation*, Acta Math. Sinica, 29 (4) (1986), 544–549.

[236] A.A. Ligun, *Best approximations of differentiable functions by algebraic polynomials*, Izv. Vuzov Ser. Mat. (1980), no. 4, 53–60.

[237] J. Löfström, *Best approximation in $L_p(w)$ by algebraic polynomials*, Studia Sci. Math. Hungar, 20 (1985), 375–394.

[238] J. Löfström and J. Peetre, *Approximation theorems connected with generalized translations*, Math. Ann. 181 (1969), 255–268.

[239] G.G. Lorentz, *Bernstein Polynomials*, University Press, Toronto, 1953.

[240] G.G. Lorentz, *An unsolved problem, in On Approximation Theory*, (P.L. Butzer and J. Korevaar, eds.), p. 185, Birkhäuser Verlag, Basel, 1972.

[241] S.M. Lozinskii, *The converse of Jackson's theorem*, Dokl. Akad. Nauk SSSR, 83 (5) (1952), 645–647.

[242] A. Lupaş, *Mean value theorems for positive linear transformations*, Rev. Anal. Numér. Théor. Approx. 3(1974), 2 (1975), 121–140.

[243] A. Lupaş, *On the approximation of continuous functions*, Publ. de L'Inst. Math., Nouvelle Série, 40 (54) (1986), 73–83.

[244] A. Lupaş, *The approximation by some positive linear operators*, in Approximation Theory (Proc. Int. Dortmund Meeting on Approximation Theory), ed. M.W. Müller et al., Berlin: Akademie Verlag, 1991, 201–229.

[245] V.N. Malosemov, *Joint approximation of a function and its derivatives by algebraic polynomials*, Soviet Math. Dokl., 7 (5), 1966, 1274–1276.

[246] V.N. Malosemov, *Joint approximation of a function and its derivatives by algebraic polynomials*, Vestn. Leningr. Univ. 23, No. 1 (Ser. Mat. Mekh. Astron.) (1968), 33–41.

[247] A. Marchaud, *Sur les dérivées et sur les différences des fonctions des variables réelles*, J. Math. Pures Appl. 9 (6) (1927), 337–425.

[248] A.A. Markov, *On the limit values of integral connected with interpolation*, Collected Works, Gostekhizdat, 1948.

[249] K.K. Mathur, *On a proof of Jackson's theorem through an interpolation process*, Studia Sci. Math. Hungar., 6 (1971), 99–111.

[250] Y. Matsuoka, *On the approximation of functions by some singular integrals*, Tôhoku Math. J., 18 (1966), 13–43.

[251] T.M. Mills and A.K. Varma, *A new proof of A.F. Timan's approximation theorem*, Israel J. Math. 18 (1974), 39–44.

[252] N. Misra, *A generalization of the Freud-Sharma operators*, Acta Math. Hung. 41 (3-4) (1983), 187–194.

[253] P. Montel, *Sur les polynômes d'approximation*, Bull. Soc. Math. France, 46 (1918), 151–192.

[254] V.P. Motornyi, *Some questions of approximation in an integral metric by algebraic polynomials*, Dokl. Akad. Nauk SSSR, 172 (3) (1967), 537–540.

[255] V.P. Motornyi, *Approximation of functions by algebraic polynomials in the L_p metric*, Izv. Akad. Nauk SSSR Ser. Mat., 35 (4) (1971), 874–899.

[256] V.P. Motornyi, *Approximation of fractional-order integrals by algebraic polynomials*, Ukr. Mat. Zh., 51 (7) (1999), 940–951.

[257] V.P. Motornyi, *On exact estimates for the pointwise approximation of the classes $W^r W^w$ by algebraic polynomials*, Ukrainian Mathematical Journal, 53 (6) (2001), 916–937.

[258] V.P. Motornyi, *Approximation of certain classes of singular integrals by algebraic polynomials*, Ukr. Math. J., 53 (3) (2001), 377–394.

[259] V.P. Motornyi and O.V. Motornaya, *On asymptotic behavior of the best approximations of classes of differentiable functions by algebraic polynomials in mean*, Dokl. Math. 52 (3) (1995), 382–383.

[260] V.P. Motornyi and O.V. Motornaya, *On the best approximation of the classes $W^r H^\alpha$ by algebraic polynomials in L_1*, East J. Approx., 1 (3) (1995), 309–339.

[261] V.P. Motornyi and O.V. Motornaya, *On the best L_1-approximation by algebraic polynomials to truncated powers and to classes of functions with $L1$-bounded derivative*, Izvestiya Mathematics 63(3) (1999), 561–582.

[262] V.P. Motornyi and O.V. Motornaya, *On asymptotically exact estimates for the approximation of certain classes of functions by algebraic polynomials*, Ukrainian Mathematical Journal, 52 (1) (2000), 91–107.

[263] O.V. Motornaya, *A refinement of an asymptotic result obtained by S.M. Nikorskii*, in: Optimization of Approximation Methods, Institute of Mathematics, Ukrainian Academy of Sciences, Kiev (1992), 63–69.

[264] O.V. Motornaya, *On the asymptotic of the best approximations of differentiable functions by algebraic polynomials in the space L_1*, Ukr. Math. J., 1993, 949–953

[265] M.W. Müller, *An extension of the Freud-Popov lemma*, in: Approximation Theory III (Proc. Int. Sympos., Austin 1980, ed. by E.W. Cheney), Acad. Press, New York, 1980, 661–665.

[266] F.G. Nasibov, *The order of the best approximations of certain classes of functions*, Izv. Vyssh. Uchebn. Zaved. Mat., 3 (1969), 35–41.

[267] L.V. Nasonova, *Approximation of functions with variable smoothness in the L_p metric*, Izv. Vyssh. Uchebn. Zaved. Mat., 11 (1979), 48–54.

[268] P.G. Nevai, *Orthogonal Polynomials*, Mem. Amer. Math. Soc. 18 (1979), No. 213.

[269] P. Nevai, *Bernstein inequality in L^p, $0 < p < 1$*, J. Approx. Theory, 27 (1979), 230–243.

[270] S.M. Nikol'skii, *On interpolation and best approximation of differentiable periodic functions by trigonometric polynomials*, Izv. Akad. Nauk SSSR, Ser Math., 10 (1946), 393–410.

[271] S.M. Nikol'skii, *On the best approximation of functions satisfying Lipschitz's condition by polynomial*, Izv. Akad. Nauk SSSR, Ser. Mat., 10 (1946), 295–322.

[272] S.M. Nikolskii, *Approximation of functions in the mean by trigonometric polynomials*, Izv. Akad. Nauk SSSR, Ser. Math., 10 (1946), 207–256.

[273] S.M. Nikol'skii, *On the best asymptotic linear method of approximation of differentiable functions by polynomials*, Dokl. Akad. Nauk SSSR, 6 (2) (1949), 129–132.

[274] T.O. Omataev, *The approximation of functions that are continuous on a segment by truncated de la Vallée-Poussin sums*, Ivuz, 6 (181) (1977), 99–106.

[275] V.A. Operstein, *A characterization of smoothness in terms of approximation by algebraic polynomials in L_p*, J. Approx. Theory, 81 (1) (1995), 13–22.

[276] P. Oswald, *Some inequalities for trigonometric polynomials in the L_p metric, $0 < p < 1$*, Izv. Vyssh. Uchebn. Zaved. Mat, 7 (1976), 65–75.

[277] P. Oswald, *Ungleichungen vom Jackson-Typ für die algebraische beste Approximation in L_p*, J. Approx. Theory, (1978), 113–136.

[278] S. Pawelke, *Ein Satz von Jackenschen Typ für algebraische Polynome*, Acta Sci. Math., 33 (1972), 3-4, 323–336.

[279] J. Peetre, *A remark on Sobolev spaces. The case $0 < p < 1$*, J. Approx. Theory, 13 (1975), 218–228.

[280] E. Picard, *Traité D'Analyse*, Tome I, Gauthier-Villars, Paris, 1891.

[281] A. Pinkus, *Weierstrass and approximation theory*, J. Approx. Theory 107 (2000), 1–66.

[282] S.A. Pichugov, *Approximation of continuous functions on a segment by linear methods*, Mat. Zametki 24 (3) (1978), 343–348.

[283] A.I. Polovina, *Best approximations of continuous functions on $[-1, 1]$*, Dopovidi Akad. Nauk UkrSSR, 6 (1964),722–725.

[284] A.I. Polovina, *Approximation of functions, defined on an interval, by algebraic polynomials*, First Republican Mathematical Conference of Young Investigators, No. 2, Kiev (1965), 560–569.

[285] A.I. Polovina, *Best uniform approximation of differentiable functions by algebraic polynomials*, Izv. Vyssh. Uchebn. Zaved., 12 (1969), 76–82. FALTA

[286] A.I. Polovina, *Approximation of differentiable functions by algebraic polynomials*, Ukr. Math. J, 30 (3) (1978), 403–411.

[287] M.K. Potapov, *On Jackson type theorems in the L_p metric*, Dokl. Akad. Nauk SSSR 111 (1956), 1185–1188.

[288] M.K. Potapov, *Some inequalities for polynomials and its derivatives*, Vestnik Moskow. Univ. Ser. I Mat. Meh., 2 (1960), 10–20.

[289] M.K. Potapov, *Approximation of non periodic functions by algebraic polynomials*, Vesnik Moskow. Univ. Ser. I Mat. Meh., 4 (1960), 14–25.

[290] M.K. Potapov, *On Approximation in L_p by algebraic polynomials*, Analyses of Contemporary Problems of Constructive Function Theory, Fizmatgiz, Moscow (1961), 64–69.

[291] M.K. Potapov, *On the structural and constructive characteristics of a function class*, Tr. Mat. Inst. Steklov 131 (1973), 211–231.

[292] M.K. Potapov, *On the structural characteristic of classes of functions with a given order of best approximation*, Tr. Mat. Inst. Steklov 134, (1975), 260–277.

[293] M.K. Potapov, *Some inequalities for polynomials and its derivatives*, Vesnik Moscow. G. Univ. 5 (1977), 70–82.

[294] M.K. Potapov, *Approximation by Jacobi polynomials*, Mosc. Univ. Math. Bull. 32 (5) (1977), 56–65.

[295] M.K. Potapov, *Approximation in the C metric by algebraic polynomials*, Thesis Dokld. All Union Congress on Approximation Theory of Functions on complex domains, Ufa, 1980.

[296] M.K. Potapov, *Approximation by algebraic polynomials in an integral metric with Jacobi weight*, Mosc. Univ. Math. Bull. 38 (4) (1983), 48–57.

[297]' M.K. Potapov, *On the approximation of functions by algebraic polynomials*, Trudy Mat. Inst. Steklov., 180 (1987), 182–184.

[298] M.K. Potapov, *Coincidence of classes of functions defined by the generalized shift operator or by the order of best polynomial approximation*, Mat. Zametki, 66 (2) (1999), 242–257.

[299] M.K. Potapov, *Approximation of functions characterized by one nonsymmetric operator of generalized translation*, Proceedings of the Steklov Institute of Mathematics, 227 (1999), 237–253

[300] M.K. Potapov, *Polynomial approximation of functions defined on a finite interval of the real line*, Fundamentalnaya i prikladnaya matematika, 6 (3) (2000), 859–871.

[301] M.K. Potapov, *Properties of a family of operators of generalized translation with applications to approximation theory*, Math. Notes, 69 (3) (2001), 373–386.

[302] M.K. Potapov, *Direct and Inverse Theorems of Approximation Theory for the mth Generalized Modulus of Smoothness*, Proceedings of the Steklov Institute of Mathematics, 232 (2001), 281–289

[303] M.K. Potapov, *Approximation of differentiable functions in the uniform metric*, Tr. Mat. Inst. Steklova, 248 (2005), 223–236.

[304] M.K. Potapov, F.M. Berisha, *On the connection between a r-order modulus of smoothness and the best approximation by algebraic polynomials*, Fundam. Prikl. Mat., 5 (1999), 563–587.

[305] M.K. Potapov, F.M. Berisha, *Direct and inverse theorems of approximation theory for a generalized modulus of smoothness*, Analysis Mathematica, 25 (1999), 187–203.

[306] M.K. Potapov and F.M. Berisha, *On the connection between the best Approximation by algebraic polynomials and the modulus of smoothness of order r*, Journal of Mathematical Sciences, 155 (1) (2008), 153–169.

[307] M.K. Potapov, M. Berisha, and F. Berisha, *On polynomial approximation in the integral metric with the Jacobi weight*, Mosc. Univ. Math. Bull. 45 (5) (1990), 31–35.

[308] M.K. Potapov and V.M. Fedorov, *On Jackson theorems for the generalized modulus of smoothness*, Trudy Mat. Inst. Steklov., 172 (1985), 291–298.

[309] M.K. Potapov, V.M. Fedorov and A. Fraguela, *A cerca de la mejor aproximacióon de funciones integrables en espacio con peso*, Rev. Cienc. Mat., 2 (1981), 19–42.

[310] M.K. Potapov and G.N. Kazimirov, *Polynomial approximation of functions with given order of the kth generalized modulus of smoothness*, Mathematical Notes, 63 (3) (1998), 374–383.

[311] S.Z. Rafalson, *The approximation of functions by Fourier-Jacobi sums*, Izv. Vyssh. Uchebn. Zaved. Mat, 4 (1968), 54–62.

[312] S. Rafalson, *An extremal relation of the theory of approximation of functions by algebraic polynomials*, J. Approx. Theory, 110 (2) (2001), 146–170.

[313] T.V. Rodina, *Two theorems on the approximation of functions that are continuous on an interval by interpolation polynomials*, Izv. Vyssh. Uchebn. Zaved. Mat., 1 (1973), 82–90.

[314] J.A. Roulier, *Best approximation to functions with restricted derivatives*, J. Approximation Theory, 17, (1976), 344–347.

[315] P.O. Runck, *Bemerkungen zu den Approximationssätzen von Jackson und Jackson-Timan, Abstract Spaces and Approximation*, Birkhäuser Verlag, Basel, 1969, 303–308.

[316] P.O. Runck and H.F. Sinwel, *On constant of approximation in the theorems of Jackson and Jackson-Timan*, in Approximation Theory III, E.W. Cheney ed., Academic Press, 1980, 763–768.

[317] P. Runck and P. Vértesi, *Some good point system for derivatives of Lagrange interpolatory operators*, Acta Math. Hung., 49 (1990), 337–342.

[318] Runge, *Über die Darstellung willkürlicher Functionen*, Acta Mathematica, 7 (1885), 387–392.

[319] Runge, *Zur Theorie der eindeutigen analytischen Functionen*, Acta Mathematica, v. 6 (1885), 229–244 and 236–237.

[320] K.V. Runovskii, *On families of linear polynomials operators in L_p-spaces*, $0 < p < 1$, Russian Acad. Sci. Sb. Math, 78 (1994), 165–173.

[321] K.V. Runovskii, *On approximation by families of linear polynomials operators in L_p-spaces*, $0 < p < 1$, Russian Acad. Sci. Sb. Math, 82 (1995), 441–459.

[322] M. Sallay, *Über ein Interpolationverfahren*, Publ. Math Inst. Hung. Acad. Sci., Vol. IX, Series A, Fast. 3 (1964-65), 607–615.

[323] R.B. Saxena, *On polynomiul of interpolation*, Studia Sci. Math. Hungar., 2(1967), 167–183.

[324] R.B. Saxena, *The approximation of continuous functions by interpolatory polynomials*, Proc. Math. Inst. Bulgar. Akad. Sci. 12 (1970), 97–105.

[325] R.B. Saxena, *A new proof of S.A. Telyakowskii's theorem on the approximation of continuous functions by algebraic polynomials*, Stud. Sci. Math. Hungar., 7 (1972), 3–9.

[326] R.B. Saxena, *Approximation of continuous functions by polynomials Freud's interpolatory polynomials*, Studia Sci. Math. Hungar. 8 (1973), 437–446.

[327] R.B. Saxena and K.B. Srivastava, *On interpolation operators. I: A proof of Jackson's theorem for differentiable functions*, Math., Rev. Anal. Numér. Théor. Approximation, 7 (1978), 211–223.

[328] R.B. Saxena and K.B. Srivastava, *On interpolation operators. II: A proof of Timan's theorem for differentiable functions*, Math., Rev. Anal. Numér. Théor. Approx., 8 (1979), 215–227.

[329] R.B. Saxena and K.B. Srivastava, *On interpolation operators. III: A proof of Teliakovskii-Gopangauz's theorem for differentiable functions*, Math., Rev. Anal. Numér. Théor. Approx., 10 (1981), 247–262.

[330] K. Scherer and H.J. Wagner, *An equivalence theorem on best approximation of continuous functions by algebraic polynomials*, Appl. Anal. 2 (1972/73), 343–354.

[331] Bl. Sendov, *On the constants of H. Whitney*, C. R. Acad. Bulgare Sci. 35 (1982), 431–434.

[332] Bl. Sendov, *The constants of H. Whitney are bounded*, C. R. Acad. Bulgare Sci. 38 (1985), 1299–1302.

[333] Bl. Sendov, *On a theorem of H. Whitney*, Dokl. Akad. Nauk SSSR 291 (1986), 1296–1300.

[334] B.Sendov and V. Popov, *Averaged Moduli of Smoothness*, Wiley, New York, 1988.

[335] Bl. Sendov and M. Takev, *The theorem of Whitney for integral norm*, C. R. Acad. Bulgare Sci. 39 (1986), 35–38.

[336] L.Ya. Shalashova, *Timan's approximation theorem for functions having a continuous derivative of fractional order*, Sov. Math., Dokl., 10 (1969), 1307–1308.

[337] L.Ya. Shalashova, *Mean approximation of functions on a finite interval*, Siberian 1972, 467–473.

[338] I.A. Shevchuk, *On the uniform approximation of functions on an interval*, Math. Notes, (1986), 521–528.

[339] I.A. Shevchuk, *Properties of the Dzyadyk polynomial kernels on a segment*, Ukr. Math. J., 1989, 457–461.

[340] I.A. Shevchuk, *Polynomial Approximation and Traces of Continuous Functions on a Segment*, Kiev, Naukova Dumka, (1992).

[341] W.G. Simon, *A formula of polynomial interpolation*, Annals Math., 2 (19) (1918), 242–245.

[342] A.S. Shvedov, *Jackson's theorems in L_p, $0 < p < 1$, for algebraic polynomials and orders of comonotone approximations*, Math. Notes, Math. Notes 25 (1979), 57–63.

[343] H.F. Sinwel, *Uniform approximation of differentiable functions by algebraic polynomials*, J. Approx. Theory 32 (1981), 1–8.

[344] K.B. Srivastava, *Proof of Jackson's theorem for differentiable functions. II*, Acta Math. Acad. Sci. Hung., (1981) 38, 15–18.

[345] K.B. Srivastava, *A remark on Mathur's paper: Simple proofs of Teliakovskii-Gopengauz's theorem*, Stud. Sci. Math. Hung. 20 (1985), 223–235.

[346] E.L. Stark, *The kernel of Fejér-Korovkin: a basic tool in the constructive theory of functions*, Functions, Series, Operators, Proc. int. Conf., Budapest 1980, Vol. II, Colloq. Math. Soc. János Bolyai 35, (1983), 1095–1123.

[347] S.B. Stechkin, *On the order of best approximations of continuous functions*, Izv. Akad. Nauk SSSR, Ser. Mat., 15, 3, 219–242 (1951).

[348] R.L. Stens, *Approximation stetiger Funktionen durch algebraische Polynome und ihre Charakterisierung durch gewichtete Lipschitzbedingungen*, Habilitatiosschrift, RWTH Aachen, 1981.

[349] R.L. Stens, *Characterization of best algebraic approximation by weighted moduli of continuity*, in Functions, Series, Operators, Vol. II, B. Sz. Nagy and J. Szabados, eds., North-Holland, Amsterdam, 1983, 1125–1131.

[350] R.L. Stens and M. Wehrens, *Legendre transform methods and best algebraic approximation*, Comment. Math. Prace Mat. 21 (1979), 351–380.

[351] A.I. Stepanets and R.V Poliakov, *Approximation of continuous functions by regular polynomials*, Ukranian 1968, 176–184.

[352] E.A. Storozenko, *Embedding theorems and best approximations*, Math. USSR Sb. **26** (1975), 213–224.

[353] E.A. Storozhenko, *Approximation by algebraic polynomials of functions of class L^p, $0 < p < 1$*, Math. USSR Izv., 11 (3) (1977), 613–623.

[354] E.A. Storozhenko, *Approximation by algebraic polynomials in L_p, $0 < p < 1$*, Vestn. Mosk. Gos. Univ., Ser. Mat. Mekh., 4 (1978), 87–92.

[355] E.A. Storonzenko and Yu.V. Kryakin, *Whitney's theorem in the L_p-metric, $0 < p < \infty$*, Sbornik: Mathematics, 186 (3) (1995), 435–445.

[356] E.A. Storonzenko, V.G. Krotov and P. Oswald, *Direct and converse theorems of Jackson type in L_p spaces, $0 < p < 1$*, Math. USSR Sbornik, 3, 27 (1975), 355–374.

[357] E.A. Storonzenko and P. Oswald, *Moduli of smoothness and best approximation in the spaces L^p, $0 < p < 1$*, Analysis Mathematica, 3 (1977), 141–150.

[358] E.A. Storonzenko and P. Oswald, *The Jackson theorem in the spaces $L^p(R^k)$, $0 < p < 1$*, Siberian Math. J., 19 (1978), 630–640.

[359] L.I. Strukov and A.F. Timan, *Sharpening and generalization of certain theorems on approximation by S.N. Bernstein's polynomials*, in: The Theory of Approximation of Functions, Nauka, Moscow 1977, 338–340.

[360] G. Sunouchi, *Derivates of a polynomial of best approximation*, Jber. Deutsch. Math.-Verein. 70, (1968), 165–166.

[361] J. Szabados, *On an interpolatory analogon of the de la Vallée-Poussin means*, Studia Sci. Math. Hungar. 9 (1974), 187–190.

[362] J. Szabados, *On a problem of R. DeVore*, Acta Math. Hungar., 27 (1-2), 1976, 219–223.

[363] J. Szabados, *On some convergent interpolatory polynomials*, in: Fourier Analysis and Approximation Theory, Coll. Math. Soc. János Bolyai, Vol. 19 (Budapest, 1976), 805–815.

[364] J. Szabados, *On the convergence of the derivatives of projection operators*, Analysis Math., 7 (1987), 349–357.

[365] J. Szabados, *Discrete Linear Interpolatory Operators*, Surveys in Approximation Theory 53, Volume 2, 2006, 53–60.

[366] J. Szabados and P. Vértesi, *Interpolation of Functions*, World Scientific, 1990.

[367] G.T. Tachev, *A direct theorem for the best algebraic approximation in $L_p[-1,1]$, $(0 < p < 1)$*, Math. Balkanica 4 (1990), 381–390.

[368] G.T. Tachev, *A converse theorem for the best algebraic approximation in $L_p[-1,1]$, $(0 < p < 1)$*, Serdica 17 (1991), 161–166.

[369] G.T. Tachev, *A note on two moduli of smoothness*, J. Approx. Theory, 81 (1995), 136–140.

[370] S.K. Tankaeva, *The Jackson theorem on the segment $[-1,1]$ and on the semiaxis $[0,\infty)$*, Izv. Akad. Nauk Rep. Kazajtan, 5 (1992), 45–49.

[371] S.A. Teliakovskii, *Two theorems on the approximation of functions by algebraic polynomials*, Mat. Sb., 70 (1966), 252–265.

[372] V.N. Temlyakov, *Approximation of functions from the Lipschitz class by algebraic polynomials*, Mat. Zametki 29 (1981), 597–602.

[373] A.F. Timan, *Approximation of functions satisfying a Lipschitz condition by ordinary polynomials*, Dokl. Akad. Nauk SSSR, 77, No. 6, 969–972 (1951).

[374] A.F. Timan, *A generalization of some results of A.N. Kolmogorov and S.M. Nikol'skii*, Dokl. Akad. Nauk SSSR 81 (1951), 509–511.

[375] A.F. Timan, *A strengthening of the Jackson theorem on best approximation of continuous functions by polynomials on a finite segment of the real axis*, Dokl. Akad. Nauk SSSR, 78 (1) (1951), 17–20.

[376] A.F. Timan, *Converse theorems of the constructive theory of functions defined on a finite segment of the real axis*, Dokl. Akad. Nauk SSSR 116. No. 5 (1957), 762–765.

[377] A.F. Timan, *Note on a theorem of S.M. Nikol'skii*, Uspehi Math. Nauk, 3 (75) (1957), 225–227.

[378] A F. Timan, *On the best approximation of differentiable functions by algebraic polynomials on a finite segment of the real axis*, Izvestia, 22 (1958), 355–360.

[379] A.F. Timan, *Theory of Approximation of Functions of Real Variable*, Pergamon Press, 1963.

[380] A.F. Timan and V.K. Dzyakyk, *On the best approximation of quasi-smooth functions by ordinary polynomials*, Dokl. Akad. Nauk SSSR, 75 (1950), 499–502.

[381] M.F. Timan, *Converse theorems of the constructive theory of functions in the spaces L_p*, Mat. Sborn. 46 (88) (1958), 125–132.

[382] M.F. Timan, *On Jackson's theorem in L_p spaces*, Ukrain. Mat. Zh. 18 (1) (1966), 134–137.

[383] V. Totik, *Problems and solutions concerning Kantorovich operators*, J. Approx. Theory, 4 (1983), 51–68.

[384] V. Totik, *An interpolation and its applications to positive operators*, Pacific J. Math., 111 (1984), 447–481.

[385] V. Totik, *Some properties of a new kind of modulus of smoothness*, Z. Anal. Anwendungen, 3 (2) (1984), 167–178.

[386] V. Totik, *The necessity of a new kind of moduli of smoothness*, Anniversary Volume on Approximation Theory and Functional Analysis, eds. P.L. Butzer, R.L. Stens and B. Sz-Nagy, ISNM 65, Birkhäuser, Basel, 1984, 233–249.

[387] V. Totik, *Sharp converse theorem of L_p polynomial approximation*, Constr. Approx. 4 (1988), 419–433.

[388] R.M. Trigub, *Approximation of functions by polynomials with integral coefficients*, Izv. Akad. Nauk SSSR, Ser. Mat., 26 (1962), 261–280.

[389] R.M. Trigub, *Characterization of integer Lipschitz classes on a segment by the rate of polynomial approximation*, Theory of Functions, Functional Analysis and their Applications, Kharkov, 1973, pp. 63–70.

[390] R.M. Trigub, *Direct theorems on approximation of smooth functions by algebraic polynomials on a segment*, Math. Notes, 54 (6) (1993), 1261–1266.

[391] T. Tunc, *On Whitney constants for differentiable functions*, Methods of Functional Analysis and Topology, 13 (2007), 95–100.

[392] A.K. Varma, *Interpolation polynomials which give best order of approximation among continuously differentiable functions of an arbitrary fixed order on $[-1,1]$*, Trans. Amer. Math. Soc., 200, 1974, 419–426.

[393] A.K. Varma, *A new proof of A.F. Timan's approximation theorem, II*, J. Approx. Theory 18 (1976), 57–62.

[394] A.K. Varma, *On an interpolation process of S.N. Bernstein*, Acta Math. Acad. Sci. Hung., 31 (1-2) (1978), 81–87.

[395] A.K. Varma and T.M. Mills, *A new proof of Telyakovski's Theorem on approximation of functions*, Studia Sci. Math. Hungar. 14 (1979), 241–256.

[396] A.K. Varma and Xiang Ming Yu, *Pointwise estimates for an interpolation process of S.N. Bernstein*, J. Austral. Math. Soc. (Series A) 51 (1991), 284–299.

[397] P. Vértesi, *Interpolatory proof of Jackson theorem*, Mat. Lapok, 18 (1967), 83–92.

[398] P. Vértesi, *On a problem of J. Szabados*, Acta Mathematica Hungarica, 28 (1-2), 1976, 139–143.

[399] P. Vértesi, *Simultaneous approximation by interpolating polynomials*, Acta Math. Acad. Sci. Hungar., 31 (3-4) (1978), 287–294.

[400] P. Vértesi, *Convergent interpolatory processes for arbitrary systems of nodes*, Acta Math. Acad Sci. Hungar., 33 (1979), 223–234.

[401] V. Volterra, *Sul principio di Dirichlet*, Rend. Circ. Mat. Palermo 11 (1897), 83–86.

[402] J.L. Walsh and T.S. Motzkin, *Polynomials of best approximation on an interval*, Proc. Nat. Acad. Sci. USA, 45 (1959), 1523–1528.

[403] J.L. Walsh and T.S. Motzkin, *Best approximation within a linear family on an interval*, Proc. Nat. Acad. Sci. USA, 46 (1960), 1225–1233.

[404] J.L. Walsh and T.S. Motzkin, *Polynomials of best approximation on an interval II*, Proc. Nat. Acad. Sci. USA, 48 (1962), 1533–1537.

[405] J.L. Walsh and T.S. Motzkin, *Mean approximation on an interval for an exponent less than one*, Tran. Ame. Math. Soc., (1966), 443–460.

[406] Weierstrass, *Über die analytische Darstellbarkeit sogenannter willkürlicher Functionen einer reellen Veränderlichen*, Sitzungsber. Akad. Berlin, 1885, 633–639, 789–805.

[407] H. Whitney, *On functions with bounded nth differences*, J. Math. Pures Appl. (9) 36 (1957), 67–95.

[408] H. Whitney, *On bounded functions with bounded nth differences*, Proc. Amer. Math. Soc. 10 (1959), 480–481.

[409] J. Wimp, *Computation with recurrence relations*, Pitman, Boston, 1984.

[410] X.M. Yu, *Pointwise estimate for algebraic polynomial approximation*, Approx. Theory Appl., 1(3), 1985, 109–114.

[411] X.M. Yu, *Pointwise estimates for convex polynomial approximation*, Approx. Theory Appl., 1 (1985), 65–74.

[412] T.F. Xie, *On two problems of Hasson*, Approx. Theory Appl., 2 (1985), 137–144.

[413] T. Xie and X. Zhou, *A Modification of Lagrange Interpolation*, Acta Math. Hung., 92 (4) (2001), 285–297.

[414] Yuan Xue-gang and Wang De-hui, *Approximation to continuous functions by a kind of interpolation polynomials*, Northeast. Math. J. 17 (1) (2001), 39–44.

[415] M. Zamansky, *Classes de saturation de certains procédés d'approximation des séries de Fourier des fonctions continues et applications à quelques problèmes d'approximation*, Ann. Sci. Ecole Norm. Sup. 66, (1949), 19–93.

[416] O.D. Zhelnov, *Whitney constants are bounded by 1 for k = 5, 6, 7*, East J. Approx. 8 (2002), 1–14.

[417] G.V. Zhidkov, *Constructive characterization of a class of nonperiodic functions*, Sov. Math., Dokl. 7 (1966), 1036–1040.

[418] S.P. Zhou, *On simultaneous approximation of a function and its derivatives*, Dokl. Akad. Nauk BSSR, 32 (6), 1988, 493–495.

[419] Laiyi Zhu, *On the modified Lagrange interpolation polynomial*, Acta Math. Sinica, 1 (1993), 136–144.

[420] A. Zygmund, *Smooth functions*, Duke Math. J., 12 (1) (1945), 47–76.

[421] A. Zygmund, *A remark on the integral modulus of continuity*, Univ. Nac. Tucuman Rev. Ser. A 7 (1950), 259–269.

Index

Class of functions
$A_p^{(r+\alpha)}$, 57
$C^{r,w}[-1,1]$, 104
$H_p^{(r+\alpha)}$, 57
H_k^{φ}, 22
$K(r,\alpha)$, 46
$L_{p,\alpha,\beta}[-1,1]$, 67
$L_{p,\alpha}[-1,1]$, 67
$S(r,\alpha)$, 46
V_1^r, 105
$W^r H_k^{\varphi}$, 22
$W_p^r[-1,1]$, 60, 103
$W^{(r)} H_p^w$, 57
$W^{(r)} A_p^w$, 57
$W^{r+1}(M,[0,2\pi])$, 9
$\mathcal{W}[0,2\pi]$, 5
Lipschitz, 2, 27
Zygmund, 6, 13, 32
Chebyshev-Fourier
   coefficient, 73

Favard's constant, 5

Inequalities
Bernstein, 3, 5

Modulus
generalized modulus, 69
generalized translation, 69
Ivanov τ moduli, 77
of continuity, 4, 68
of order r, 4
of the type of order k, 22
Potapov, 68

Kernels
Bohman-Zheng, 172
Fejér-Korovkin, 173

Jackson-de la Vallée-Poussin, 173
Jackson-Matsuoka, 16
Korovkin, 172

Operators
Bernstein, 133
Bojanic, 141
boolean sum, 162
Butzer-Stens, 142
Chawla, 141
Chebyshev, 54
DeVore, 140
Dzyadyk, 19
Fejér's means, 143
Fejér-Korovkin, 144
Landau, 139
Saxena-Srivastava, 153
Srivastava, 150
Varma, 155
Vértesi I, 152
Vértesi II, 152

Polynomials
Chebyshev first kind, 141
Chebyshev first type, 115
Chebyshev second type, 116
fundamental Lagrange interpolation
   polynomials, 115
Jacobi, 145
Legendre, 69, 140, 141, 145
zeros of Chebyshev, 115

Systems
Bernstein-Erdös type, 132
well approximating, 132